TEN FACES OF THE UNIVERSE

GOD'S UNIVERSE
THE PHYSICIST'S UNIVERSE
THE MATHEMATICIAN'S UNIVERSE
THE ASTROPHYSICIST'S UNIVERSE
THE EXPANDING UNIVERSE
THE ORIGIN OF THE UNIVERSE
NOBODY'S UNIVERSE
THE GEOPHYSICIST'S UNIVERSE
THE BIOLOGIST'S UNIVERSE
EVERYMAN'S UNIVERSE

TEN FACES OF THE UNIVERSE

FRED HOYLE
California Institute of Technology

W. H. FREEMAN AND COMPANY
San Francisco

Library of Congress Cataloging in Publication Data

Hoyle, Fred, Sir.
Ten faces of the universe.

Includes index.
1. Cosmology. I. Title.
QB981.H756 523.1 76-44336
ISBN 0-7167-0384-X
ISBN 0-7167-0383-1 pbk.

Printed in the United States of America

1 2 3 4 5 6 7 8 9

CONTENTS

PREFACE

To the reader, the preface of a book is made to seem the beginning, but to the author—to most authors, at any rate—the preface is the end, giving a chance to comment finally on what has been written already. So it is here.

My plan of this book called for bringing together several distinct views of the nature of things, of the universe, as a play is fashioned from a set of characters. My characters proved for the most part to be reasonably manageable in their allotted roles, but one caused me trouble, the last one of all. The trouble with him was that he tended to hog the stage, so much so that it might have been better to give him a whole play to himself, as indeed other authors have sometimes done. But I have written about this particular fellow before, indeed, almost a quarter of a century ago.* So by now I have become rather impatient with his tantrums.

*In *A Decade of Decision,* London: Heinemann, 1953.

Other authors have been content to treat the world population problem more gently than I have done in my last chapter, more gently than the problem deserves, by taking the overly kind view that, with the spread of knowledge to the underdeveloped four-fifths of the world, a demographic transition will occur, whereby populations at last become stabilized. The trouble with this view is that it schedules stabilization to occur in the twenty-first century, by which time the world count of people will have risen to a monstrous 15 billion. I have never believed in this supposed demographic transition, because I do not believe the world has enough food, energy, and materials, or that these resources can be developed at a rate fast enough, to support such a population. Nor do I understand how its adherents can reconcile the contradiction between their views and the fact that, as the years pass, the multitudinous poor become poorer still. And ominously, too, the political systems of the wealthy minority, instead of becoming stronger and more innovative, are becoming more feeble.

The position is not one of unremitting gloom, however. Some ten to fifteen years ago a momentous step was taken. For the first time in human history a psychologically acceptable technology for controlling birth rates became available. By use of this technology, world populations could be stabilized right now. There is hence no reason why we should be compelled to wait on the uncertainties of a demographic transition. Indeed, since nine-tenths of today's social problems already arise from overpopulation, there is every reason not to permit the threefold or fourfold further expansion which the adherents of demographic transition are unwisely prepared to accept.

If it be asked why stabilization is not taking place right now, the answer lies squarely with the ineptitude of world political systems. Governments cannot be blamed, as they sometimes are, for failing to apply unknown forms of technology. But once a crucial technology exists, failure to avoid certain disaster by applying it becomes a crushing political failure. On this, I believe it is necessary to take an unyielding position, as I have done here in my final chapter.

Doing so gives mankind the best chance to avoid a dark catastrophe in the not too distant future.

As always, I wish to acknowledge the great help that my wife Barbara has given me in assembling the material for this book and in the writing of it.

Fred Hoyle

Cockley Moor
May 1976

TEN FACES
OF THE UNIVERSE

1

GOD'S UNIVERSE

ÉVARISTE GALOIS
(1811–1832)

1

GOD'S UNIVERSE

In the year 1930 James Jeans gave the Rede Lecture in the University of Cambridge. The lecture, considerably expanded in scope, appeared in print shortly thereafter under the title *The Mysterious Universe.* The book made a considerable impact on my schoolboy mind (I had still three more years to go at grammar school). The highpoint of the lecture lay in its implication that God is a mathematician. Jeans led up to this mighty assertion in a delightfully quaint fashion:

"Fifty years ago, when there was much discussion on the problem of communicating with Mars, it was desired to notify the supposed Martians that thinking beings existed on the planet Earth, but the difficulty was to find a language understood by both parties. The suggestion was made that the most suitable language was that of pure mathematics; it was proposed to light chains of bonfires in the Sahara, to form a diagram illustrating the famous theorem of Pythagoras, that the squares of the two smaller sides of a right-angled triangle are together equal to the square on the greatest side. To most of the inhabitants of Mars such signals would convey no meaning, but it was argued that mathematicians on Mars, if such existed, would surely recognize them as the handiwork of mathematicians on Earth. . . . So it is with the signals from the outer world of reality. . . . We have already considered with disfavor the possibility of the universe having been planned by a biologist or an engineer; from the intrinsic evidence of his creation, the Great Architect of the universe now begins to appear as a pure mathematician."

Jeans soon found himself in the midst of an uproar. He was denounced from the pulpits; philosophers wrote books designed to expose the threadbare quality of his reasoning; and, worst of all, the British scientific establishment began to consider him just a bit of an outsider. To his credit, Jeans stuck to his guns throughout this assault, and, in the eyes of one schoolboy at least, he gave as good as he got in the ensuing controversy.

His opponents seemed unable to understand that, although Jeans himself had a tolerable debating position, the attributes of God so frequently and confidently announced from the pulpit were quite indefensible:

> God, the father—i.e., the family man;
> God, the maker of all things—i.e.,
>   a craftsman or artisan;
> God almighty—a war leader;
> God in heaven, wherever that might be.

All such attributes, being plainly man-made, contribute nothing of value to the concept of God. It was these man-made attributes which the critics should have been attacking.

We might seek to attack the concept of God the mathematician by arguing that mathematics is also a man-made attribute. But, to put debating points aside, there is much more to it than that. From the concepts of mathematical structure and elegance, a *predictable* quality emerges in the world. The story of the weak interaction, (to be described in Chapter 3) provides us with an example here. From the concepts of mathematical structure, it was found possible, some twenty years ago, to predict successfully the outcome of certain critical experiments that have since been performed, experiments that can in principle be performed by *anybody,* and which, when performed correctly, always give the *same* results. The veracity of the individual does not enter the matter at all, whereas it does when one is confronted by so-called "revealed truths."

Throughout the history of science, similar situations have occurred, so often, in fact, that we can be sure that real truths do exist, truths that are mathematical in form. Scientists always tackle their unsolved problems in the belief that such truths are there to be found, and so far they have never been disappointed. The only verifiable statements, verifiable by *anybody,* which can be made about the world are those which rest on these mathematical truths.

Opposition to this point of view usually springs from the feeling that it somehow downgrades the importance of the individual. It is therefore of interest to notice that an unusual way can be found to reassert the importance of the individual, a way with a certain peculiar validity, as we shall see in Chapter 6. There is a school of philosophical idealism, more popular formerly than it is today, going back to Bishop Berkeley, who said:

> All the choir of heaven and furniture of Earth, in a word, all those bodies which compose the mighty frame of the world, have not any substance without the mind. . . . So long as they are not actually perceived by me, or do not exist in my mind, or that of any other created spirit, they must either have no existence at all, or else subsist in the mind of some eternal Spirit.

If we assume that he referred to eternal Spirit as a concession to his cloth, then Berkeley was saying that the world only exists because *I* or *you* perceive it. On this view, one might attempt to argue that the mathematical structure which the physicist finds present in the universe is a product of his own mode of perception—that the human brain, being mathematical in its perceptions, imprints those perceptions on the universe. The objection to this argument is that the human brain is *not* highly mathematical—we learn mathematics only with great effort. There are many other aspects of our brains which come more readily to us, and these we do *not* find mirrored in the structure of the universe.

This Berkeleyan-style argument appears to me to reverse the order of the horse and cart. We humans have emerged from a long process of competitive biological evolution. In a competitive evolutionary struggle, the creature that best understands the nature of its environment is most likely to win, as we humans—so far—have won. Understanding the environment implies understanding the rules which determine cause and effect in the environment, and if these rules are mathematical in form, then the understanding of the environment implies a perception of mathematics. This, I would say, is why humans possess some degree of mathematical talent. If we equate God with the universe, then the biblical pronouncement that God has created Man in His own image becomes applicable to these ideas in a somewhat strange way.

But equating God with the universe is something most people are not at all happy to do. Most people prefer to conceive of God as being outside the universe. Looking back at the quote from Jeans, we can see that this was his mode of thought: there is to be a Great Architect standing outside His own mathematical creation. The objection to the concept of God as being outside the universe is that nothing sensible can be made of it. We are to give no attribute to God except as a thinker of mathematical laws—but, in doing so, we cast God in Man's image, for it is *we* who have to think in order to perceive the mathematical laws. All other attributes of God are without meaning, and it does far more harm than good to go on playing around with them.

As an example, consider the fact that something approaching the conditions of a civil war exists today in Northern Ireland. Although secular problems have played a significant role in causing this situation, a religious quarrel between Protestants and Catholics is generally conceded to lie at the root of it. Since the Christian religion is supposed to be based on an ethic of "love thy neighbor," this quarrel is perverse and contradictory. A while ago, I happened to suggest in a talk that a quick and simple solution to the Irish problem would be to arrest every priest and clergyman in Ireland and to commit every man jack of them to long jail sentences on the charge of causing civil war. When the ensuing laughter subsided, I was surprised that not a single person among the fair-sized company present seemed to doubt that this odd-sounding proposal would in the long run solve the Irish problem.

Priests and clergymen do not intend to cause pain, but when they persist in repeating nonsense words and concepts to children, and insist that those words and concepts have great hidden significance, they *do* cause pain. The mental frustration of it all then erupts into violence, when two groups of people, fed on different nonsense words, intermingle with each other. Where the Irish have a sensible objective, such as defeating England at rugby, nobody cares who is Catholic and who Protestant. Together they simply get on with the job, and they do it very well. There is no such thing as Catholic eyes, or Protestant legs, or Marxist numbers, or capitalist geometry. Combining a nonsense word with a valid word always produces this kind of ridiculous association.

Man has been imbued with religious impulses, certainly for the last 5,000 years, and probably for very much longer than that. Are these impulses all to be suppressed? What significance shall be attached, for example, to the impulse of worship? Worship covers the whole gamut from the absurd to the serious, from the admirable to the not so admirable. It is not so admirable to sacrifice an animal to a god in order to secure an advantage for oneself. When the women adorn the church with flowers, this is admirable. When the first families sat down to the first Thanksgiving dinner, that too was admirable. When a scientist spends a good fraction of his life trying to discover the mathematical form of some new physical law, knowing all the while that the chance of personal success is not high, that is worship. When a government spends money on an accelerator or a telescope, that is worship. Whenever anyone, at whatever level of sophistication, makes the effort to understand a little more about the world, that is worship. There will never be any long-term purpose for our species other than understanding of the universe. If this purpose does not prove sufficient for us, if we are impelled to invent all manner of nonsensical substitutes, then very likely we shall not survive as the dominant animal on the Earth for very much longer.

It is a major tragedy that the rich treasure house of physical laws, to be discussed in general terms in Chapters 2 and 3, is not accessible in detail to everyone. The sharp clarity of perception becomes lost when only a general description can be given. The trouble lies in the mathematics. Mathematics, more than any other subject of study, turns people off. That this should be so, in spite of the many billions of dollars spent annually on education, is not only one of the grievous scandals of our age, but also rather puzzling, since everybody starts life as a potentially competent mathematician.

Watch a baby between six and nine months old, and you will observe the basic concepts of geometry being learned. Once the baby has mastered the idea that space is three-dimensional,* it

*Well-skilled in yowling for its food, the baby knows about time from the beginning.

reaches out and begins grasping various kinds of objects. It is then, from perhaps nine to fifteen months, that the concepts of sets and numbers are formed. So far, so good. But now an ominous development takes place. The nerve fibers in the brain insulate themselves in such a way that the baby begins to hear sounds very precisely. Soon it picks up language, and it is then brought into direct communication with adults. From this point on, it is usually downhill all the way for mathematics, because the child now becomes exposed to all the nonsense words and beliefs of the community into which it has been so unfortunate as to be born. Nature, having done very well by the child to this point, having permitted it the luxury of thinking for itself for eighteen months, now abandons it to the arbitrary conventions and beliefs of society. But at least the child knows something of geometry and numbers, and it will always retain some memory of the early halcyon days, no matter what vicissitudes it may suffer later on. The main reservoir of mathematical talent in any society is thus possessed by children who are about two years old, children who have just learned to speak fluently.

Modern advanced societies have found no way to make effective use of the marvelous years in a child's life from two to five. Quite apart from the mathematical talent awaiting to flower, the child has uncanny linguistic ability. It is well within the capacity of the average child to learn four or five languages, perfect in accent and syntax. Yet we let the child fritter away these priceless years, feeding itself on the arbitrary conventions of our society. Only at a much later age, when the linguistic ability has largely been lost, is the child expected to learn a new language. Results, achieved at great expense to taxpayers, are by then poor.

The child, the sparkling brilliance of the first eighteen months now long forgotten, is entered into school, where it begins the long wearisome climb from grade to grade. It is at this point, three to four years too late, that its teachers seek to rekindle the dead embers of its mathematical talent. For the most part the fire never comes alight again, and the teachers—desperate to awaken interest and attention—try all manner of devices, devices like the so-called "new" math, which strangle the talent of those few children who had somehow managed to retain the inventive qualities of the earliest years. So it comes about that a society essentially devoid of mathematical skill is produced.

Having taught mathematics for twenty years myself, at the level of the dreaded and famous Cambridge Tripos, I have strong opinions on this subject. In the first place, students should never be *taught* mathematics at all. Everybody should learn individually, because each person has a different pace. Setting the pace right is critical, because all important ideas must be clearly and completely learned, to a point where ideas and techniques become wholly instinctive. A slow pace does not matter very much, because there is ample time in life to become an expert mathematician, almost regardless of pace. What does matter, crucially, is for the learning to be so precise and complete that returning over old ground is scarcely ever necessary. It is just because students attempt to go too fast, and are then forced into endlessly reviewing old material, that so many of them fall by the wayside.

Now, how is the student to learn for himself? By solving puzzles. The functions of the teacher should be, first, to select in a wise way the material on which the puzzles are based, second, to make sure the puzzles are well-suited in difficulty to the sophistication of the student, third, to answer questions, and finally, if the teacher is capable of it, to give an occasional word of inspiration. A wisely set puzzle is one which naturally suggests an important mathematical idea. Wisely selected puzzles taken as a whole should cover the range of mathematical ideas appropriate to the student's particular stage of development, and they should do so neither too narrowly nor too widely. The teacher has also the important task of deciding the sequence of development. Experienced teachers know only too well that ideas at very different levels of difficulty should not be presented in step. Each idea must be adjusted to the capacity of the student, so that he will be able to see it for himself. When this happens, when "the light dawns," the student is far less likely to forget than if he were being lectured in class by the teacher. It is also important to notice here that solving puzzles is exactly how a baby learns throughout those important first eighteen months. By learning in this way, the student is continuing the natural process which was performed so successfully in the earliest months.

Looking back over my own early years, I realize now how very fortunate I was. With my father conscripted into the British Army in the second year of the first World War, just after I was born, my mother was very alone and therefore able to give much attention to my education. She taught me the multiplication tables before I was three, not as a matter of understanding, but purely by rote. This she did by jogging me on her knee, pretending I was riding a horse. Meanwhile we chanted the multiplication tables together. By the time I was four I knew what the tables meant, and I could write them down in proper arabic notation. Then I was sent to school at the age of five. The headmistress must have been told of my interest in numbers, for she immediately began to teach me herself. To my astonishment she wrote the numbers all in roman form. The first day I took this to be some kind of joke, but when the performance was repeated from day to day, I realized the old girl really knew no better. At the end of the first week, a small child of five walked determinedly out of that school, never to darken its doors again.

The local town had a population of about 20,000, quite a big place for a little chap of five to be walking about every day for a couple of months. Somehow I managed to convince my mother that I was still at school. Instead I visited the wool mills. Nobody, as I recall, seemed to take the slightest notice of a small boy wandering earnestly among the clicking looms. But my favorite place was beside the lock gate on the local canal. I watched endlessly as the boats came through, backward and forward, carrying all manner of goods. I listened carefully to the rich language of the bargees. Thus equipped, first with the multiplication tables, and now with a wide understanding of the English language, it was inevitable that I should subsequently have had an easy passage through life.

Years later, at the age of nine, my class at school was still chanting those tables, so that the weaker members could learn them by rote as I had done so many years before. To relieve the tedium of the daily chant, a few of us orchestrated the performance. We stationed ourselves at different places in the class, leading our respective sections in counterpoint. The performance drove the teachers well-nigh crazy, but they accepted it, perhaps because jobs were scarce in the mid-nineteen twenties. The wildness of it all encouraged the weaker spirits, to the extent that I do not think any one among them failed to master the tables by the age of ten, a better performance, I believe, than is usual nowadays.

The excruciating boredom of those arithmetic classes, and—perhaps worse—of classes in the English language, led me to start thinking in parallel, a faculty which has continued down the years, a faculty which sometimes offends my friends greatly. One day we had a language class on the use of the article, in the course of which we were required to make up examples, and then to read our creations aloud to the rest of the class. Results were ludicrous. When a close friend read out "an pig," I said to myself: If we hadn't had this lesson, Johnny would never have written anything as absurd as "an pig." There and then I resolved to have nothing further to do with such classes. I have kept to this resolve. Never once in my life have I consulted a book on the usage of the English language—only a dictionary occasionally, to check the conventional spelling of some word. In writing, I struggle to express my thoughts clearly, never "correctly"—whatever that might mean.

By thus dismissing a whole batch of classes as useless, I had plenty of time at school to think about all manner of interesting things. It was utterly necessary, however, to avoid the heinous crime of "not paying attention." In my day, you inevitably suffered physical punishment for not paying attention. Stupidly doing so, by staring out of a window, at some interesting cloud formation, for instance, would inevitably earn for you a stinging series of blows about the head. To a little chap like me, such blows were not to be thought of, so I avoided staring at interesting things in the wide world outside the windows. Even this precaution was not sufficient, however, because teachers would sometimes ask you to repeat what they had just said. So long as you were able to do so, it was accepted that indeed you were really "paying attention." So I learned, always at all times, to let the last few words the teacher was saying register in my brain. Each new sentence wiped out the preceding one, thereby permitting instant recall of the new sentence. This invention won for me hour upon hour of glorious contemplation.

I mention these personal details because I believe they cast some light on the mysterious death of the French mathematician Évariste Galois. Following a happy childhood in the countryside, Galois was sent to a miserable school in Paris. He detested all classes except mathematics, evidently learning to occupy himself by working mathematical problems in his head. He became so proficient at "mental mathematics" that he hardly ever troubled to write in the formal way mathematicians normally use. His slender notes were examined after his death by the rather conservative and established mathematician Joseph Liouville. Liouville wrote: "My zeal was well rewarded, and I experienced an intense pleasure at the moment when, after filling in some slight gaps, I saw the complete correctness of the method by which Galois proves, in particular, the following beautiful theorem. . . ." Mozart composed much of his music in the same way.

Entry to the École Polytechnique and to the Ecole Normale was barred to the young Galois, since, knowing little except mathematics, he failed to satisfy the entrance requirements. A special appeal was made on the ground of his exceptional mathematical talent, but this failed too, for Galois, when called to an interview, failed to explain his ideas to the assembled professors.

In the political atmosphere of the Paris of 1830, Galois became a revolutionary. It seems he achieved some notoriety in this role, by threatening death to the King Louis-Philippe, in the course of an after-dinner speech. There is little evidence that his activities constituted any actual threat to the King. Nevertheless, he was eventually imprisoned, along with a number of his revolutionary friends. He was released after six months incarceration. Then, within a short time of his release, Galois received a challenge to a duel, which he felt unable to refuse, even though he seems to have had little doubt of the result. The night before, he sat up late, committing his last mathematical results to paper with breakneck speed. In the early hours of the following day, in a field on the outskirts of Paris, he was shot and left for dead by his adversary. Picked up by a casual passerby and taken by cart to a Paris hospital, he died the following morning. Like Mozart, he was buried in a common grave.

Such, then, are the bare bones of the story of the life and death of Évariste Galois. The classical biography* of Galois, in an attempt to add flesh to these bones, suggests that he was done to death by his royalist enemies, as does E. T. Bell in his book *Men of Mathematics.* There are dark hints that the release from prison was but a device for encompassing his death, a necessary preliminary to his being matched against a highly skilled assailant in royalist pay. But why would Galois feel it critical to his honor that he should accept the challenge of a right-wing agent, especially if the agent were a

---

*P. Dupuy, "La Vie d'Évariste Galois," *Annales de l'École Normale,* série 3, **13,** 197–266.

known marksman? Gallic logic suggests, on account of a girl, mentioned in a letter from Galois to a friend, written shortly before his death:

> "I beg patriots and my friends not to reproach me for dying otherwise than for my country. I die the victim of an infamous coquette. It is in a miserable brawl that my life is extinguished. . . . Pardon those who killed me, they are of good faith."

Why should Galois have asked "pardon" for a right-wing opponent, and why should he speak of such a person as being of good faith? From this letter it is clear that Galois was killed by his revolutionary friends, an interpretation confirmed by a second letter:

> "I have been challenged by two patriots—it was impossible for me to refuse. I beg your pardon for having advised neither of you. But my opponents had put me on my honor not to warn any patriot. Your task is very simple: prove that I fought in spite of myself, that is to say, having exhausted every means of accomodation."

It is possible that the "infamous coquette" was the source of a purely personal quarrel, but it is the normal biological rule among mammals that sexual quarrels between two males cease as soon as one side seeks "accomodation." It is the normal rule that either party to such fights can simply walk away, which is just what Galois seems to have attempted to do.

The more likely possibility is that Galois' habit of working mathematical problems in his head, his ability to think in parallel, caused serious personal animosities, and perhaps suspicions, to develop during the six months of imprisonment. There may have been suspicions that Galois was not wholly for the "cause," or even that he was an *agent provocateur.* In a city only recently emerged from the Terror, a city of informers and counterinformers, such an accusation would not seem farfetched, particularly when brought against a man so deeply engrossed in his own private thoughts.

Near the end, Galois was visited by a priest come to administer the last rites. "Are you come to mock me?" he asked. And to his younger brother, who cried as he sat on the bed, Galois remarked, "Don't cry. I need all my courage to die at twenty." The scrawled pages he wrote on the eve of the duel have been described as Galois' "time-outlasting memorial." What he had done was to make basic discoveries in the part of mathematics known as the theory of groups.

The theory of groups underlies the whole of modern physics.

FURTHER READINGS

E. T. Bell, *Men of Mathematics.* Simon and Schuster, 1937.

J. H. Jeans, *The Mysterious Universe.* Cambridge University Press, 1930.

2

THE PHYSICIST'S UNIVERSE

MURRAY GELL-MANN
(1929-   )

GEORGE ZWEIG
(1937-   )

2

THE PHYSICIST'S UNIVERSE

"TWA Flight X from London to Chicago is now ready for boarding at Gate Y."

I have forgotten the gate and the flight number, but I shall never forget one singular aspect of that journey—I saw the Sun rise in the west. The plane flew out over the Atlantic Ocean, heading to the northwest, the short January afternoon drawing to its close in the usual way, with the Sun sinking below the southwestern horizon. The external world darkened. People around me slept, talked, gulped cocktails, listened with headphones to the music of their choice. But then in a slight and subtle way the quality of the light in the cabin began to change. To my astonishment, the glow in the western sky was getting brighter, not fainter as it was supposed to. And the sky just went on and on growing brighter, until, miraculously it seemed, the fierce golden disc of the Sun appeared again on the horizon.

Fortunately for my sanity, by this time I had worked out what had happened. Our route had taken us so far north, over the southern part of Greenland, that the speed of the plane was overcompensating the spin of the Earth. It was like being on a planet that rotated east-to-west, instead of our everyday west-to-east. With the coming of supersonic commercial flights, this phenomenon will eventually become well-known, but for a while I had a distinctly queasy feeling about it. Not so my fellow passengers. As I recall, nobody else seemed even faintly curious. My fellow passengers simply went on talking, drinking, and listening, stuffed up (as I thought) by the headphones and by the monotonous thud of the music of their choice. I have come since then to realize that different people see such a journey in very different ways. Some see it as a bore to be endured, some (especially in the first class section of the cabin) as a kaleidescope of food and drink, some as the beginning of a new life—I well remember the anxious faces of a group of immigrants on their way to Australia. Each human journey has a color of its own, sometimes rich color, sometimes a little drab, but color nonetheless. In this, we differ in our personal lives from the physicist when he is expressing himself technically.

The physicist does not trouble himself with headphones, with pretty girls serving food and drink, with the air crew or the passengers. The physicist lumps everything together, plane and all, into a single object which he calls a "particle," the motion of which he displays in the manner of Figure 2.1, with time plotted one way and distance the other way. This graphic method of displaying the journey is known in physics as a *spacetime diagram.*

Actual flight paths are not quite straight, of course, but this additional detail causes no difficulty, since we could generalize Figure 2.1 to a more realistic form in which two space dimensions are used as well as the time dimension. The two space dimensions could be used to show the variation of latitude and longitude along

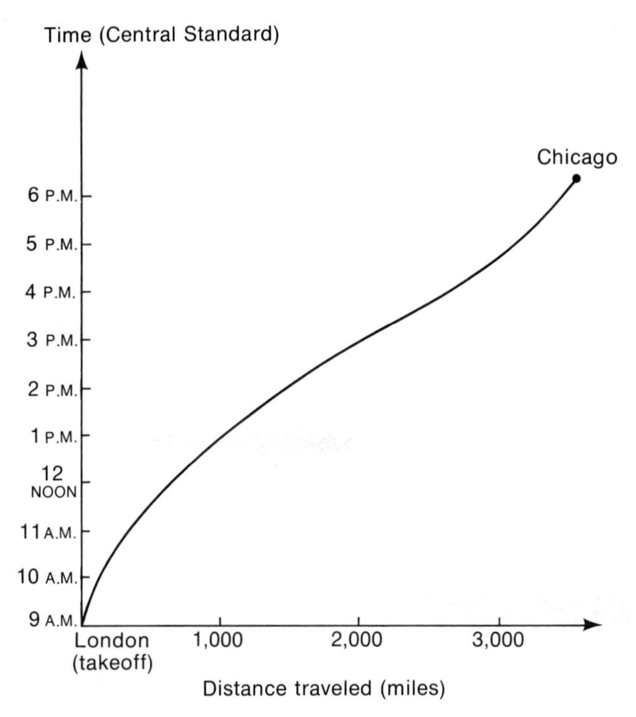

FIGURE 2.1
The time-distance plot for an airplane going from London to Chicago. Note that the plot is not a straight line, because the plane goes more slowly at the beginning and at the end than it does in midflight.

the path of the aircraft. And if we wish to show the variations in height of the plane, we could add a third dimension of space; our complete spacetime diagram would then have three dimensions of space and one dimension of time. So we pass from the simple concept of Figure 2.1 (one time dimension, one space dimension) to the more complex idea of a four-dimensional diagram (one time dimension, three space dimensions). According to the physicist, the whole drama of the universe is enacted in such a four-dimensional spacetime diagram.

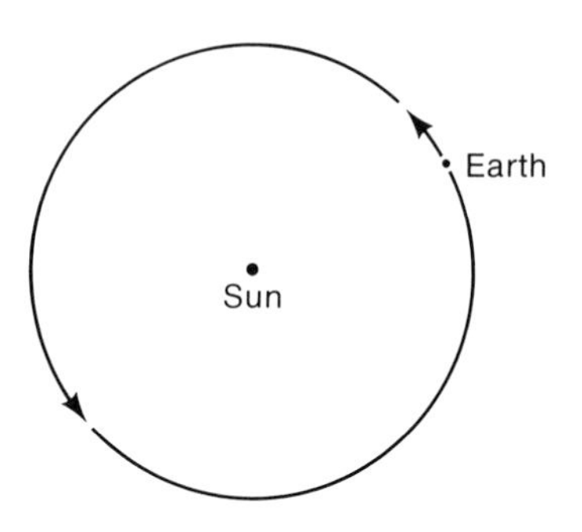

FIGURE 2.2
A purely spatial diagram of the orbit of the Earth around the Sun. At any explicit moment of time, the Earth is at a particular place in the orbit. As time goes on, the Earth moves in the manner indicated.

It is important to notice the difference between a spacetime diagram and a purely spatial diagram, a difference illustrated by Figures 2.2 and 2.3. In Figure 2.2 we have a purely spatial diagram, showing the Earth's annual orbit around the Sun. The dot representing the Earth moves around the Sun as "time goes along." But such a way of thinking is rather vague, because it does not show how the spatial motion is related to the passage of time. If we wish to display the passage of time explicitly, as in Figure 2.3, the orbit of the Earth in the resulting spacetime diagram must be shown as a spiral having the Sun as its axis.

Here already we run into a curious problem. We tend to prefer Figure 2.2 to Figure 2.3, because we attach subjective importance to explicit moments of time. We think of Figure 2.2 as an explicit moment, with other orbital configurations of the Earth and Sun occurring at other explicit moments. In Figure 2.3, on the other hand, we have the *whole history* of the Earth's movement, without any one configuration being singled out as having special significance. There is nothing in Figure 2.3 related to the concept of the *present moment of time*. In physics as it is normally understood, there is no explicit moment denoting the "present." All moments of time exist together, with the whole world occupying a four-dimensional spacetime diagram.

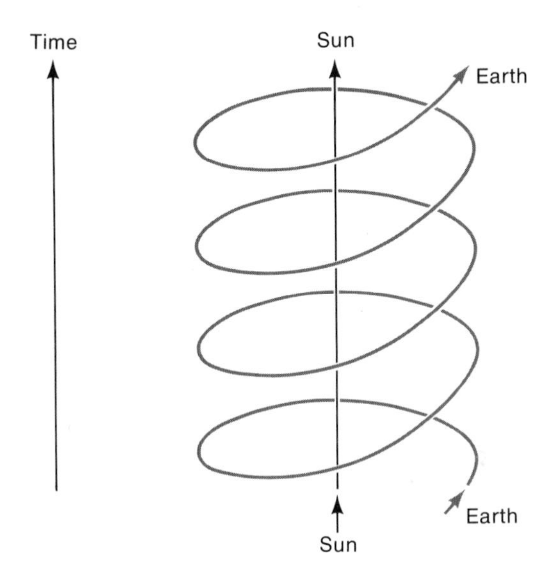

FIGURE 2.3
When the time dimension is added to Figure 2.2, the path of the Earth becomes a spiral. There is no representation of the "present moment" in this diagram; all times exist together.

It can be argued that the physicist gains clarity at the cost of leaving out interesting detail. Figure 2.1 gives nothing of the make-up of the "particle" which journeys from London to Chicago. The headphones, the cocktails, the air crew and the passengers with their colorful (or not so colorful) lives, have all been lost. To this, the physicist will respond by saying that conventional descriptions achieve little, because the details which they give simply *do not go far enough*. What are headphones? What is a cocktail? How does the human brain function? The physicist looks for answers to these further questions, and if we press him for details, he is only too likely to reply with an avalanche of information.

Bodies like an airplane or like the Earth are composite. That is to say, they can be divided into smaller pieces. How small? Without giving a clear-cut answer to this question, the ancient Greeks believed there had to be an ultimate limit to the smallness of the pieces into which any body can be divided, and they referred to these ultimate pieces as atoms. Scientists of the nineteenth century held the same belief, and they set about trying to find out how many different kinds of atoms there are in the world. In this enquiry they were reasonably successful, as will be seen from Table 2.1, which gives a full list of all the atoms now known, together with their cosmic abundances and with the year in which each of them was discovered.

Separating bodies into their constituent atoms was an important problem in the nineteenth century. Scientists learned techniques for preparing standard samples of the different kinds of atoms, standard in the sense that each sample contained the same number of atoms, irrespective of the kind of atom—a sample of hydrogen contained the same number of hydrogen atoms as there were carbon atoms in a sample of carbon, for example. With this done, they could compare the weights of the different samples, and so calculate the relative weights of the different atoms. Except for the neighboring pairs, cobalt and nickel, and tellurium and iodine, Table 2.1 is arranged according to increasing weight. Hydrogen is the lightest atom, helium the next lightest, and so on. Taking the weight of a hydrogen atom to be 1, they found the weight of a helium atom to be about 4, that of a carbon atom to be about 12, oxygen 16, aluminum 27, iron 56, and so on. Such relative weights were then referred to as *masses*.

The mass of a large body is simply the mass of its constituent atoms. Imagine the body divided into atoms, and proceed to separate the atoms into their various kinds. Count the number of atoms of each kind, allowing for the fact that different atoms have

TABLE 2.1
The elements

Z	*Name*	*Chemical symbol*	*Date of discovery*	*Abundance in cosmic material*[a]
1	Hydrogen	H	1766	3.18×10^{10}
2	Helium	He	1895	2.21×10^{9}
3	Lithium	Li	1817	49.5
4	Beryllium	Be	1798	0.81
5	Boron	B	1808	350
6	Carbon	C	Old	1.18×10^{7}
7	Nitrogen	N	1772	3.64×10^{6}
8	Oxygen	O	1774	2.14×10^{7}
9	Fluorine	F	1771	2,450
10	Neon	Ne	1898	3.44×10^{6}
11	Sodium	Na	1807	6.0×10^{4}
12	Magnesium	Mg	1755	1.06×10^{6}
13	Aluminum	Al	1827	8.5×10^{5}
14	Silicon	Si	1823	10^{6}
15	Phosphorus	P	1669	9,600
16	Sulfur	S	Old	5.0×10^{5}
17	Chlorine	Cl	1774	5,700
18	Argon	A	1894	1.17×10^{5}
19	Potassium	K	1807	4,205
20	Calcium	Ca	1808	7.2×10^{4}
21	Scandium	Sc	1879	35
22	Titanium	Ti	1791	2,770
23	Vanadium	V	1830	262
24	Chromium	Cr	1797	1.27×10^{4}
25	Manganese	Mn	1774	9,300
26	Iron	Fe	Old	8.3×10^{5}
27	Cobalt	Co	1735	2,210
28	Nickel	Ni	1751	4.8×10^{4}
29	Copper	Cu	Old	540
30	Zinc	Zn	1746	1,245
31	Gallium	Ga	1875	48
32	Germanium	Ge	1886	115
33	Arsenic	As	Old	6.6
34	Selenium	Se	1817	67
35	Bromine	Br	1826	13.5

TABLE 2.1 (*continued*)

Z	*Name*	*Chemical symbol*	*Date of discovery*	*Abundance in cosmic material*[a]
36	Krypton	Kr	1898	47
37	Rubidium	Rb	1861	5.88
38	Strontium	Sr	1790	26.8
39	Yttrium	Y	1794	4.8
40	Zirconium	Zr	1789	28
41	Niobium	Nb	1801	1.4
42	Molybdenum	Mo	1778	4
43	Technetium	Tc	1937	unstable
44	Ruthenium	Ru	1844	1.9
45	Rhodium	Rh	1803	0.4
46	Palladium	Pd	1803	1.3
47	Silver	Ag	Old	0.45
48	Cadmium	Cd	1817	1.42
49	Indium	In	1863	0.189
50	Tin	Sn	Old	3.59
51	Antimony	Sb	Old	0.316
52	Tellurium	Te	1782	6.41
53	Iodine	I	1811	1.09
54	Xenon	Xe	1898	5.39
55	Cesium	Cs	1860	0.387
56	Barium	Ba	1808	4.80
57	Lanthanum	La	1839	0.445
58	Cerium	Ce	1803	1.18
59	Praseodymium	Pr	1879	0.149
60	Neodymium	Nd	1885	0.779
61	Promethium	Pm	1947	unstable
62	Samarium	Sm	1879	0.227
63	Europium	Eu	1896	0.085
64	Gadolinium	Gd	1880	0.297
65	Terbium	Tb	1843	0.055
66	Dysprosium	Dy	1886	0.351
67	Holmium	Ho	1879	0.079
68	Erbium	Er	1843	0.225
69	Thulium	Tm	1879	0.034
70	Ytterbium	Yb	1878	0.216

TABLE 2.1 (*continued*)

Z	*Name*	*Chemical symbol*	*Date of discovery*	*Abundance in cosmic material*[a]
71	Lutetium	Lu	1907	0.0362
72	Hafnium	Hf	1923	0.210
73	Tantalum	Ta	1802	0.0210
74	Tungsten	W	1781	0.160
75	Rhenium	Re	1925	0.0526
76	Osmium	Os	1803	0.745
77	Iridium	Ir	1803	0.717
78	Platinum	Pt	1735	1.40
79	Gold	Au	Old	0.202
80	Mercury	Hg	Old	0.40
81	Thallium	Tl	1861	0.192
82	Lead	Pb	Old	4.0
83	Bismuth	Bi	1753	0.143
84	Polonium	Po	1898	unstable
85	Astatine	At	1940	unstable
86	Radon	Rn	1900	unstable
87	Francium	Fr	1939	unstable
88	Radium	Ra	1898	unstable
89	Actinium	Ac	1899	unstable
90	Thorium	Th	1828	0.058
91	Protoactinium	Pa	1917	unstable
92	Uranium	U	1789	0.0262
93	Neptunium	Np	1940	unstable
94	Plutonium	Pu	1940	unstable
95	Americium	Am	1945	unstable
96	Curium	Cm	1944	unstable
97	Berkelium	Bk	1950	unstable
98	Californium	Cf	1950	unstable
99	Einsteinium	Es	1955	unstable
100	Fermium	Fm	1955	unstable
101	Mendelevium	Md	1955	unstable
102	Nobelium	No	1958	unstable
103	Lawrencium	Lw	1961	unstable

[a]Abundances from a recent compilation by A. G. W. Cameron (*Space Science Reviews,* **15** (1970), 121–146). Notice that the abundances are *relative* to each other, with 10^6 for Si taken as the standard of reference.

different masses. Work always in terms of hydrogen, the atom of least mass. Thus for hydrogen count 1, but for each carbon atom count 12, because each carbon atom has twelve times the mass of a hydrogen atom. For each oxygen atom, count 16, and so on. At the end of this counting process, you then have a measure of the mass of the body in question, reckoned in terms of the hydrogen atom as your standard unit. You can imagine this process being carried out for any body, for any planet, or for any star.

Coming back for a moment to our spacetime diagram for a body like an airplane, we can achieve greater precision by thinking of the airplane not as a single path in the diagram, but as a bundle of paths, with a separate path for each atom. We can think of such a bundle as a cable formed from a large number of fine threads.

There is no need to draw a separate diagram for each different body. Each body can be represented in the same diagram by its own cable. Sometimes a thread will emerge out of a cable, as when an atom escapes from the Sun into the wind of atoms which the Sun emits all the time. Sometimes a thread will emerge from one cable and, after wandering by itself for a while, will enter another cable, as when an atom in the solar wind enters the atmosphere of the Earth. Although the concepts have not been changed very much, the apparently simple picture of Figure 2.1 has suddenly become quite complex. The cable representing the airplane contains on the order of 10^{31} threads, so small are the atoms compared to the much larger bodies of our everyday world.

Yet we are still very far from being done with our original question: How small are the pieces into which matter can ultimately be divided? One achievement of physics in the first half of this century was to prove that the atoms of Table 2.1 do not constitute the answer to this question, for atoms are already composite structures, built from more basic particles. Most of the weight of an atom lies in a small central region which contains two kinds of particles, *protons* and *neutrons*. Outside the nucleus are much lighter particles called *electrons.* In a normal atom the numbers of electrons and protons are equal. The number of neutrons is not always fixed, however, even for atoms of the same kind. When there are alternatives for the number of neutrons present in a specific kind of atom, the alternatives are called *isotopes.* Thus the atom of chlorine has two isotopes, one with 17 protons and 18 neutrons, the other with 17 protons and 20 neutrons. Each kind of atom has a definite number of protons, however, usually denoted by Z. It is worth noticing that in Table 2.1, we classified the atoms in terms of the increase in Z, with the proton number increasing by one at each step.

Looking at atoms in this way, we need only three particles to describe our physics, instead of the many kinds of atoms which were all thought to be basic in the nineteenth century. However, although we thereby gain a better understanding of the nature of atoms, our spacetime cables of threads become still more complex; an atom, instead of being represented by a single thread, has itself become a bundle of threads, with a separate thread for each electron, proton, and neutron. The threads for the protons and neutrons are glued together to form the nucleus of the atom, but the threads for the electrons weave patterns that extend far outside the nucleus. To understand the difference in scale, think of the nucleus of an atom (say, an atom of oxygen) as being the size of a golf ball; then the patterns woven by the electron threads would occupy a baseball park. The study of these electron patterns forms the science of atomic physics, whereas the study of the compact proton and neutron threads forms the science of *nuclear physics.*

It is possible to build "artificial atoms," atoms whose nuclei contain numbers of protons and neutrons that are different from the numbers that occur naturally in atoms. Artificial atoms have the curious property that either some of their protons change into neutrons, or some of their neutrons into protons, thereby altering the atom itself. Such unstable atoms are said to be radioactive. The presence of unstable atoms in the debris of nuclear bombs, and in the waste products of nuclear reactors, constitutes a hazard against which careful protection must be provided. The reason for the hazard is that other particles—fast-moving electrons, in particular—are also generated whenever there is an interchange of neutrons and protons. If fast-moving electrons enter the human body in large numbers, they cause serious physiological damage, and sometimes even death.

The fact that neutrons and protons are interchangeable according to the formulas

$$\text{neutron} \rightarrow \text{proton} + \text{other particles}$$

or

$$\text{proton} \rightarrow \text{neutron} + \text{other particles}$$

suggests that protons and neutrons are not really different particles, but instead different manifestations of some more fundamental entity. The same suggestion has come from the discovery of further new particles. From about 1940 onward, more and more of these new particles were found. Six of them were given the rather odd designations of Λ, Σ^+, Σ^0, Σ^-, Ξ^-, Ξ^0. These six, together with the proton (p) and the neutron (n), which the six resemble in some ways, form a family of eight particles, called *baryons*—a circumstance which physicists have labeled the "eightfold way."

The other six members of this eightfold family are not found in the atoms of our everyday world, because each of the six changes into either a neutron or a proton, according to the formula

$$\Lambda, \Sigma^+, \Sigma^0, \Sigma^-, \Xi^-, \Xi^0 \rightarrow \text{n or p} + \text{other particles},$$

in an exceedingly small interval of time, on the general order of 10^{-10} seconds.* The "other particles" involved in these transitions

*From an everyday point of view it is remarkable that experiments can be carried out within time intervals as small as this. Highly sophisticated equipment is needed for such experiments. This is one of the reasons why modern physics has become a financially expensive study.

are not like p, n, Λ, Σ^+, Σ^0, Σ^-, Ξ^-, Ξ^0. They are either electrons or particles like electrons. These "other particles" are called *leptons,* and they are quite distinct from the eightfold family. (For one thing, every member of the eightfold family is of greater mass than the leptons.) Yet there are indeed other baryon families akin to the eightfold family. There is a family of ten, of which one member, the Ω^-, is rather famous, because its properties were predicted ahead of time, before its existence was confirmed by experiments in the laboratory.

The idea that permitted the existence of Ω^- to be predicted ahead of time came in the early 1960's. It was suggested, independently by M. Gell-Mann and by G. Zweig, that all these heavy particles are composite. The eightfold and the tenfold families can be built out of three more basic particles, which have become known as *quarks.* The three different quarks are sometimes referred to as the "up" quark, the "down" quark, and the "strange" quark. The members of the eight and tenfold families are all obtained by combining quarks three at a time. Thus the proton is a combination of 2 "up" quarks and 1 "down" quark; the neutron is 1 "up" quark and 2 "down" quarks; the Λ is a combination of 1 "up" quark, 1 "down" quark, and 1 "strange" quark. The Ω^- is a combination of 3 "strange" quarks. The rules governing all these combinations are those of the theory of groups—the same theory that, in its basics, was discovered by Galois.

The use of peculiar words to describe these new particles is deliberate. All words have associated meanings, and if a physicist wishes to avoid everyday associations, he chooses a new and hitherto unfamiliar word like "quark." On the other hand, the word "strange" does have well-known associations, and this again is deliberate. The strange quark is not present in either the proton or the neutron, the two members of the eightfold family that are found in our everyday world. Particles containing the strange quark are all evanescent, belonging to the "strange" new world revealed by modern experiments.

At the time I am writing, two further particles, different from the members of the previously known families, have just been discovered. To explain them, physicists are currently suggesting that there may be a fourth kind of quark, called a "charmed" quark. By calculating what will happen if a charmed quark is combined with two of the others (up, down, and strange) according to the rules of the theory of groups, physicists are currently predicting the existence of a whole new array of particles. The big question is whether the predicted particles will actually be found. If they are, physicists will be convinced that, however esoteric their theories may have become, a satisfactory measure of truth lies within them.

Let us look once again to our spacetime picture of cables and threads. We formerly had a thread for each neutron and each proton in the nucleus of every atom. Now each neutron and each proton must itself be represented by a bundle of three quarks. We have thus arrived at an exceedingly intricate tapestry of cables and threads—the avalanche of detail that I promised you. As the physicist delves further and further into the detailed structure of the universe, new and more intricate levels of complexity are revealed.* Can we expect an ultimate level to be reached, an ultimate stage beyond which nothing remains to be discovered? I think not. Let me explain why.

Suppose scientists hold a certain theory (about something or other) which we denote by *T*. For the theory to be considered "good," it must represent a part of the observed behavior of the world to within a satisfactory tolerance. For example, Newton's theory of gravitation permits the next total eclipse of the Sun by the

*Here is an example of an even further level of complexity: it is now suspected that each species of quark has three varieties, usually referred to as "colors." The up quark thus has three "colors," as do the down, strange, and charmed quarks. This gives us an array of twelve kinds of quarks—and more new questions.

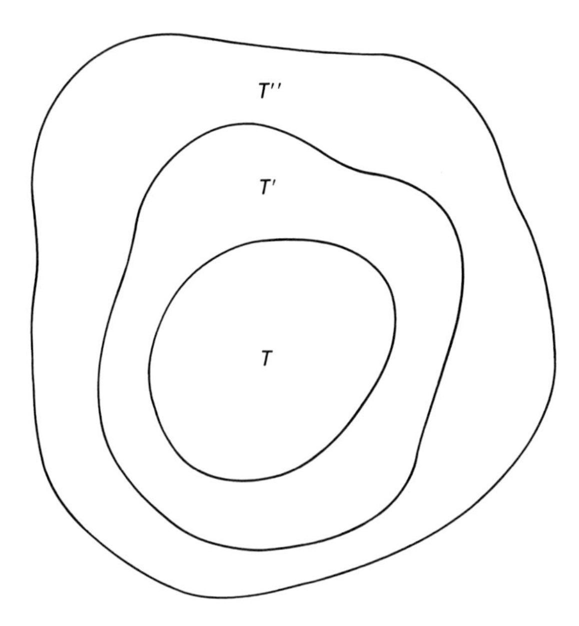

FIGURE 2.4
The territory encompassed by the theory T' encloses that of T, and so on.

Moon to be predicted ahead of time to within an accuracy of about one second. Since this tolerance is regarded as satisfactory, Newton's theory is considered to be "good" for making eclipse predictions.

If now an improved theory, T', should be discovered, we would mean that T' encompasses a wider range of experience than T, not that T is to be replaced within its own range of applicability. We can think of the situation territorially as in Figure 2.4; the territory covered by T' includes that covered by T. And if we could find a further improved theory, T'', we would mean that the territory of T'' encompasses that of T'.

Is it likely that an ultimate theory, $T'''^{\cdots}$, which fully and precisely encompasses all phenomena, will ever be discovered? My own belief is that the answer to this question is no. Whenever we widen the domain of experience, new ideas always seem to be necessary. To achieve a complete and perfect theory, we would need to widen the range of experiment and observation to include everything, the whole universe. If the answer to our question were indeed affirmative, we would then be able to describe the whole universe in terms of the processes occurring in our own brains—i.e., in terms of a small subregion of the universe itself. This would, I feel, be an implausible circumstance. I suspect the truth of the matter is that no full description of the universe is possible within only a part of the universe. Any complete theory would inevitably encompass the whole universe—our $T'''^{\cdots}$ would be the universe itself. An ultimate theory, like the Holy Grail, is something the physicist must always seek but will never find.

FURTHER READINGS

S. L. Glashow, "Quarks with Color and Flavor," *Scientific American,* **233** (October 1975), 38.

3

THE MATHEMATICIAN'S UNIVERSE

JAMES CLERK-MAXWELL
(1831–1879)

(Photo courtesy of the Cavendish Laboratory,
University of Cambridge.)

3

THE MATHEMATICIAN'S UNIVERSE

The concept we arrived at in Chapter 2, of a profusion of cables and threads in a spacetime diagram, is by no means a free-wheeling picture. The cables representing large bodies like the Earth, and the threads representing basic particles, do not go just anywhere. They are subject to important restrictions. Only certain pictures can be woven in the tapestry of the world. It is by means of these restrictions that "cause" and "effect" become related, that order and structure become established.

In Chapter 2 we saw that an airplane and its contents—the cocktails, the food, the crew and the passengers—are made up from an exceedingly large number of atoms, on the order of 10^{31}. The astonishing thing is that so enormous a quantity, 10,000,000,000,000,000,000,000,000,000,000 atoms, can organize themselves into the complex components of our everyday lives, the airplane and its structure, the seat you sit in, the book or newspaper that you might read, the conversation of a neighbor. All this is done by the restrictions that form the topic of this chapter.

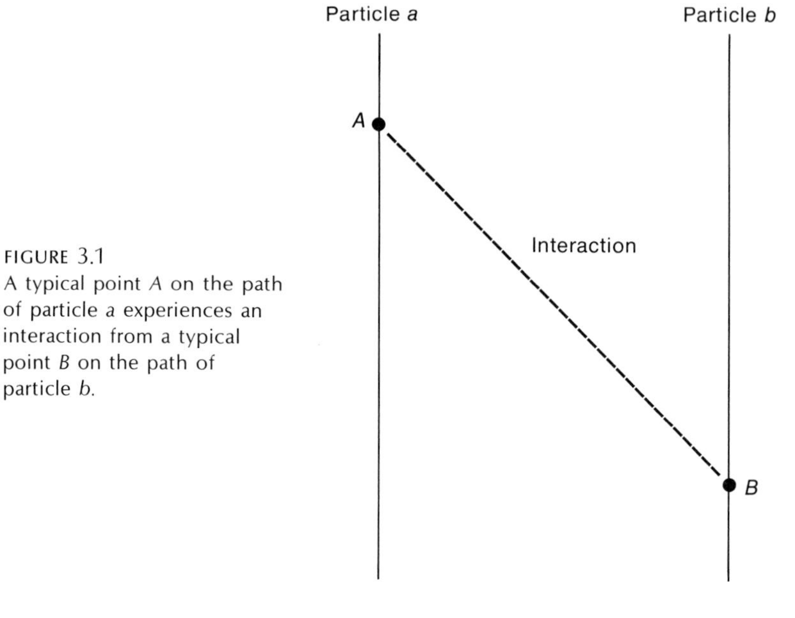

FIGURE 3.1
A typical point *A* on the path of particle *a* experiences an interaction from a typical point *B* on the path of particle *b*.

The nature of the restriction on the thread representing particle *a* is illustrated in Figure 3.1. At a typical point *A* on the path of a particle *a*, an influence—or "interaction," as it is usually called—is received from other particles; in Figure 3.1 an interaction is received from point *B* on the path of particle *b*. Such an interaction affects the form of the path of particle *a*. Thus the path or "thread" which any particle follows is influenced by its interaction with other particles. The particle paths are thus not chosen arbitrarily. The paths are all interlinked with one another, and it is the business of the mathematician (or of the physicist wearing a mathematical hat) to calculate how the linkages go and how the restrictions operate on the paths of the particles.

Four distinct forms of interaction have been discovered. They are:

(1) gravitational,
(2) electrical,
(3) weak,
(4) strong.

FIGURE 3.2
Apollo 17: Rover vehicle on the desolate lunar surface.
(Courtesy of NASA.)

The gravitational interaction is important for the large cables in our tapestry, but not so important for the fine threads. It is the gravitational interaction that holds the Earth in its orbit around the Sun. It holds us down on the surface of the Earth, and on the surface of the Moon, as in Figure 3.2.

The electrical interaction is perhaps the most widely studied of the four. It is the electrical interaction of the protons in the nucleus of an atom, acting on the surrounding electrons, which prevents the electrons from flying off and leaving the nucleus "bare." The electrical interaction also causes atoms to stick together in bunches; for example, two atoms of hydrogen and one of oxygen form a bunch written as H_2O. Such bunches are called *molecules,* and H_2O is the molecule of water. The electrical interactions between molecules of water are highly important and unusual, since they give bulk liquid water its remarkable properties as a solvent—properties that appear to be essential for the development of life. The detailed study of the properties of moderate-sized bunches of atoms forms the science of chemistry.

The weak interaction is that which permits a quark to change from one kind to another, whereas the strong interaction is what binds together the three quarks that form a proton, a neutron, or one of the other particles of the eightfold and tenfold families that we studied in Chapter 2. The strong interaction also binds together the protons and neutrons in the nucleus of an atom. Without the strong interaction, the protons and neutrons in a nucleus would simply fly apart.

The four interactions operate on very different scales. The weak and the strong interactions are important over distances of 10^{-13} to 10^{-12} centimeters, this being the scale of the nuclei of atoms. The electrical interaction is especially important on the scale of the electrons which surround the nuclei of atoms, and also on the scale of molecules, 10^{-8} to 10^{-7} centimeters. The gravitational interaction is important on the scale of large bodies like the Earth, the Sun, and the distant universe.

All four interactions are expressed mathematically, that is, by mathematical equations. All mathematical equations are of the form $x = y$, where x and y are the same number. It may be wondered why such a triviality is considered important. If x and y are the same number, then obviously $x = y$. However, x is a number calculated or measured by one procedure, and y is a number calculated or measured by another, distinct procedure. Initially, we seem to have no reason to suppose that numbers obtained in two apparently different ways will be the same. When we find they are indeed the same, we express our surprise and delight by triumphantly writing $x = y$! This is the essence of mathematics, nothing more. For example, x might be a number calculated from the geometric form of the path of particle *a* in Figure 3.1, and y might be a number calculated from the interaction coming from particle *b*. Then the statement $x = y$ relates the form of the path of particle *a* to the interaction on it. This was precisely the logic by which Newton arrived at his theory of gravitation.

Knowing the form of the path of particle *a* by this technique, we can predict ahead of time where particle *a* is going to be. Thinking of *a* as the Moon in its orbit around the Earth enables the next eclipse of the Sun to be predicted. The reader may care to enquire exactly what prediction has actually been made, and may care, when the time arrives, to check the prediction. It will assuredly be found that the prediction works. *No other way of successfully predicting the future behavior of the world has ever been found.*

It is an admirable quality of many sporting activities that, when they are played at the highest level of skill, an expert performer has to learn to lose. By this I do not mean gritting the teeth into a strained smile and muttering ill-felt congratulations to the winner. I mean trying for a win to the very last gasp of human effort, losing, and then not letting the loss affect performance in the next game. The hard thing to avoid is getting into a slump. Something of the same sort happens in science, except that scientists tend to suffer collectively rather than individually, as in a loss experienced by a whole team. Suppose an equation $x = y$ has long been used successfully for making predictions ahead of time, as with the eclipse predictions of Newton's theory. Then it often happens that all scientists will come to believe that x, obtained according to one procedure, is precisely equal to y obtained by another procedure.

Then suppose a new, more accurate method for comparing x with y becomes available, and suppose the new method shows that x is not quite equal to y—very nearly, but not quite. For practical purposes the difference may not seem important, and the temptation is to ignore it.

But history shows that, although most people do take this line of least resistance, the greatest discoveries are made by the rare person who remains dissatisfied that x is not quite equal to y. Kepler revolted against the fact that x, as calculated from his own improvements on the theory of Copernicus, did not quite equal y as measured by the observations of the Danish astronomer Tycho Brahé; here x was the predicted position of the planet Mars in its orbit around the Sun, and y was Tycho's observation of the actual position of Mars. Kepler's revolt led to the discovery of the true form of the orbit of Mars, and this discovery led to the whole of modern science—to the discoveries of Newton and to all that flowed from those discoveries.

A similar situation occurred toward the end of the nineteenth century, with x calculated from Newton's theory for a certain feature* of the orbit of the planet Mercury and with y obtained from observations analyzed by the American astronomer Simon Newcombe. The Newtonian x was very nearly equal to the observed y, but not *quite* equal. From a practical point of view the discrepancy seemed small, but it turned out to provide a crucial verification of an entirely new theory of gravitation. When Einstein formulated this new theory, it led to a slightly different value of x, which turned out triumphantly to be really equal to the observed y. So once again from a small discrepancy a great revolution of scientific thought emerged.

*The rotation of the long axis of the elliptic orbit of Mercury.

The history of the electrical interaction followed somewhat similar lines. Early successes were achieved particularly by the French scientists Charles Augustin de Coulomb (1736–1806) and André Marie Ampère (1775–1836). These successes were extended by the English scientist Michael Faraday, and his successes in turn culminated in the remarkable achievement of James Clerk Maxwell. A century earlier, the biographer James Boswell had asked the literary lion, Samuel Johnson, if Scotland did not have "noble wild prospects." Johnson replied: "Norway, too, has noble wild prospects; and Lapland is remarkable for prodigious noble wild prospects. But, Sir, let me tell you, the noblest prospect which a Scotchman ever sees is the high road that leads him from Auld Reekie to London." Maxwell, a Scotsman who followed the road from Auld Reekie (Edinburgh) to Cambridge (better than London), achieved such a discovery as one might hope would silence every literary wit for the next thousand years.

Maxwell's mother died when he was nine years old. Educated in part at school and in part by his father until the age of sixteen, Maxwell then spent three years at Edinburgh University, taking the road thereafter to Cambridge. Teaching in Cambridge at that time was in the hands of coaches of considerable technical competence, but usually with little imagination. However, Maxwell was lucky. His coach, William Hopkins, stood aside, mastering a natural inclination to force his student into the standard Cambridge mold. Troubled in his conscience when the time came for Maxwell to enter the final examinations (the dreaded and famous Cambridge Tripos), Hopkins remarked, "I have never sent a man into the examination so ill-prepared, and yet I have some confidence in him, for in natural philosophy (physics) he is incapable of thinking wrongly." Maxwell was placed second in order of merit in the examination, a tremendous performance for a supposedly ill-prepared student.

His first major postgraduate investigation was a rather dull affair on the rings of the planet Saturn. But Maxwell was led, by thinking about the many particles which make up the rings of Saturn, to consider the many particles in a hot gas. His paper on that subject provoked the exclamatory comment from the great German scientist Ludwig Boltzmann (1844–1906), "*This* is the way to do the theory of gases," much as two centuries earlier Johann Bernoulli had commented, on an anonymously printed work of Isaac Newton, "Ah, I recognize the lion by his paw."

Following his paper on the rings of Saturn, we find Maxwell writing to the scientist William Thomson, later Lord Kelvin (1824–1907), asking if Kelvin would mind if Maxwell were to devote some of his attention to the problems of electricity. (Such were the niceties of the nineteenth century.) Maxwell's decision to turn his "attention to the problems of electricity" was like the thunder before the dawn.

On the continent of Europe the most famous mathematicians, perhaps the most inventive mathematicians of all time, had already turned their attention to the problems of electricity. Not one of them had even come close to a solution, not Gauss and Riemann in Germany, not Cauchy in France. The wonder is that Maxwell, equipped only with the student mathematics he had picked up almost casually in Cambridge, solved the problem by calculating the detailed mathematical form of the electrical interaction. The work was published in 1864, under the title "A Dynamical Theory of the Electromagnetic Field."

Maxwell's work simply forced the further discoveries, made at the beginning of this century by H. A. Lorentz, H. Poincaré, A. Einstein, M. Planck, and H. Minkowski, which go by the name of the special theory of relativity. Actually, there is not very much more content in this theory beyond what Maxwell had achieved, and this additional content is mainly four equations of the kind described above, $x = y$, with x calculated from the geometrical form of the path of particle a and with y a number calculated from the electrical interaction coming from particle b. These four equations, which led to the subsequent cliché $E = mc^2$, were first given in a preliminary form by Lorentz, and later in a more complete form by Minkowski. Experience in teaching all these matters at a post-graduate level (i.e., with the relevant mathematical detail included) shows that at least a semester course is required for the student to understand what Maxwell did. Thereafter, the special theory of relativity is found to be "easy"; the average student can master the main points of it within a week. Indeed, the student is usually puzzled by why it took scientists of the late nineteenth century some thirty years to see what was almost staring them in the face.

Scientists during the nineteenth century were plagued by a wrong concept, and the real problem they faced was to get rid of this blockage. Nobody* could understand how an interaction can go from particle b to particle a in Figure 3.1 simply through space and time. They all felt that some form of material that transmitted the interaction had to be present between the points A and B of this figure. Since no such material was found by experiment, it had to be an inherently invisible form of material, an invisible jelly which vibrated when it passed the interaction from B to A. Riemann and Cauchy failed in their attempts to describe the electrical interaction, because they sought to do so in these terms. Even after Maxwell, who eschewed the jelly, everybody still thought there had to be some way in which the stuff could be brought into the story, and they spent twenty years or more in seeking to develop this wrong idea. Of course, if they had used the words "invisible jelly," the

---

*Except possibly Gauss, who seems to have failed to describe the electrical interaction for a much deeper reason than his contemporaries; the route which Gauss tried to follow does lead to the same results as Maxwell obtained, but in a more difficult way.

absurdity of what they were trying to do would soon have been recognized. But they called it "aether," which, being a Greek word of high cultural associations, was much harder to decry. This is why scientists today use nonsense words like "quark" to describe even very serious concepts. It is hard enough to avoid deceiving oneself without having confusing words like "aether" to cope with.

It was Lorentz who first suspected the error, Poincaré who first became convinced of it, and Einstein who destroyed it in the eyes of the whole scientific world in one hammer blow in the year 1905. But with the exception of the new equations of Lorentz and Minkowski, the achievement had not really been a major one. It amounted to writing the mathematical equations discovered by Maxwell in what today we would call a four-dimensional format—that is to say, with the time dimension treated like the spatial dimensions, much as we have drawn our spacetime diagrams in these chapters. The special theory of relativity was really a reorientation of scientific attitude. The great discovery had already been made thirty years earlier, by Maxwell.

Maxwell's theory shed such a fierce clear light on so many questions that the frontiers of knowledge were able to advance rapidly in the present century, so rapidly indeed that a further very deep and important problem was soon encountered. Let us look again at the equation $x = y$, with x determined by the form of the path of particle a and with y determined by the interaction from particle b; the explicit procedure for calculating y had been found by Maxwell, and the explicit procedure for calculating x had been discovered by Lorentz, Planck and Minkowski. Was x always precisely equal to y? For the substantial cables in our spacetime diagram, the answer to this question was affirmative, and it was the substantial cables that concerned scientists up to the early years of this century, for the good reason that the fine threads representing electrons, protons, neutrons, . . . , had not yet been discovered.

With the discovery of the fine threads, however, particularly the electron threads, it was found that x is *not* equal to y. So the equation $x = y$ works if a and b in Figure 3.1 are cables but not if they are threads.

This difficult and fundamental problem faced scientists in the two decades following the discovery of the special theory of relativity. The solution was found, notably by Heisenberg, Schrödinger, and Dirac, in the years between 1925 and 1930. Unlike the solutions to most major scientific problems, which are usually very much welcomed, the solution to this one was quite unwelcome to some of the most distinguished scientists of the time. The solution seemed at first sight to have been bought at too high a price; too many of the gains achieved by science during the preceding 300 years seemed to have been lost. At the beginning of this chapter, we saw that the tapestry of cables and threads in our spacetime diagram is not free-wheeling, that restrictions on it are created by interactions that lead to equations of the kind $x = y$, with the numbers x and y computed from different aspects of the tapestry. It followed that if one were forced to give up at least some of the equations used previously to restrict the tapestry, the picture woven by the cables and threads would become more free again. It was this loss of control over the picture on the tapestry which many scientists of the time disliked.

Just how free could the picture become? What the new ideas said was that any picture was possible. The reader might be inclined to exclaim: "Fine! Just what *I* like." But would you? An outfielder throws to home plate, hoping to cut off the runner. He expects the ball to rise in the air and then to fall, because of the interaction of the Earth's gravity on the ball. He hopes the ball will pitch a few yards in front of the catcher, that the catcher will sweep a hand through the ball and proceed to tag the runner. What he does *not* expect is that, after being thrown, the ball will soar up into the sky and fly through space to land on the Moon. Yet in a free tapestry this could happen. Or the Earth, instead of continuing to follow its usual orbit around the Sun, might take off like a Pioneer spacecraft and go to Jupiter. Such a world would be hard to live in.

The saving grace is that not all pictures on the tapestry are equally probable. It is possible for the baseball to soar up into the sky, but the chance of this happening is exceedingly minute, much less than 1 in 10^{100}. The outfielder would have to throw an awful lot of balls before he would hit the Moon. So the cables in the tapestry can still be taken to follow the same paths they did formerly. Other paths are possible *in principle,* but the chance of the tapestry taking a significantly different form is so exceedingly minute that we can ignore it—we have lots of more important problems to worry us.

The situation is quite different, however, for the threads within the cables. It used to be thought that the threads representing an electron in an atom followed a helical path rather like that of the Earth in Figure 2.3, although, of course, much smaller in scale. But according to the new theory, although a path of the form of Figure 2.3 is certainly possible, many other paths are available, all with essentially as good a chance of being followed. What was done between 1925 and 1930 was to discover how to calculate explicit numbers giving the chance for each of these paths to be followed. Each detailed form of the tapestry could then be given a probability. The method for making such calculations became known as *quantum mechanics.*

During the past forty years, more has been written about the philosophical implications of quantum mechanics than about any other aspect of the physical sciences. The basic problem, as most commentators see it, is that physics presents many possible tapestries differing from one another in the details of the threads which represent the particle paths. The question is asked: Is there an actual universe corresponding to each of these tapestries? Or is there just one universe, and, if so, what is meant by the other possible forms permitted by the theory? To anticipate my discussion in Chapter 7, I offer the following answers to these questions: There are *not* many universes. The actual universe follows a definite tapestry. Although our present *theory* does not permit us to distinguish the actual tapestry from other apparently possible forms, by *experiment* we can find at least parts of the actual pattern of the universe—that is, we find them simply by looking at the world. Does the definiteness thus achievable by experiment, but not by theory, imply that the present theory is incomplete? We will take up this question when Chapter 7 is reached.

The new generation of younger scientists takes a different view of quantum mechanics than did the generation of 1930. The older, prequantum theory is now seen to have been too rigid, too constrained. The universe, if it really conformed to the picture which scientists held in the first two decades of the century, would be quite markedly less interesting than the actual universe. To take just one example, the methods used by astronomers to calculate the distances of galaxies, and so to build a picture of the large-scale structure of the universe, would not be possible. The regret felt by the older scientists* over the quantum theory has passed. It is seen that Nature can be wiser than even the greatest of scientists.

The first attempt to describe the weak interaction was made by Enrico Fermi in the early 1930's. It was immediately apparent to physicists that Fermi's theory was not the only one that could be proposed. If we think again of $x = y$, with x a number computed by Fermi's proposed theory and with y a number measured by experiment, there were analogous, but different, ways in which one might proceed to compute x. Since the different ways for computing x did

*Schrödinger once remarked, "I do not like it, and I am sorry I ever had anything to do with it."

not lead to the same number, it seemed that all that was needed was an accurate practical measurement of y. Then the correct theory was the one which gave $x = y$.

At first this procedure of choice by experiment swung away from Fermi's explicit proposal, toward an alternative that many physicists felt to be rather artificial and contrived. The first experiments were then found to be incorrect; their measurements of y were wrong, and so the tide swung back to Fermi. The work continued throughout the 1940's, but without a satisfactory decision. There was, of course, not just one experiment of the type $x = y$, but many experiments that could be used in this way. No formulation of the weak interaction seemed to lead to values of x that were equal to y for all such experiments, and by the early 1950's a crisis situation was reached.

It was possible to divide the various ways of formulating the weak interaction into two broad categories, known technically as the "parity conserving" class and the "parity non-conserving" class. Nobody had ever looked very seriously at the "parity non-conserving" class, for the apparently good reason that several experiments seemed to lead to y values very different from the x values given by this whole class of possibilities. But then, around 1955, an experiment was performed by Miss C. S. Wu at Columbia University (following up a suggestion by T. D. Lee and C. N. Yang) which clearly contradicted the opposite class of "parity-conserving" possibilities. The experiment was widely repeated in many laboratories throughout the world, always with the result that the parity-conserving theories were wrong. It followed that either Fermi's original concept was at fault, or one or more of the earlier experiments was wrong.

At this point several physicists decided on a quite new approach. Rejecting the parity-conserving theories, they looked to see which parity non-conserving possibility appeared to have the simplest and most elegant mathematical structure—a criterion based on aesthetics rather than on an empirical comparison with experiment. The outcome was a certain explicit formulation of the weak interaction, a formulation that still contradicted certain earlier experiments. Were these earlier experiments right? Here was an interesting clash between the human concept of mathematical structure and the harsh reality of practical experiment. Which side would win when the experiments were repeated? The answer turned out to be the intellectual concept of elegance and structure. The earlier experiments had been wrong.

The study of the weak interaction is not yet ended. In the middle 1960's, a few physicists began to wonder if the best choice of the mathematical structure had been made after all. They introduced the new concept of a connection between the weak interaction and the electrical interaction into their arguments, and were led to a new formulation with still different experimental consequences. The new experiments were very hard to perform, and it has been only during the past two years that results have been obtained. The results, concerned with what are known technically as "neutral currents," support the new ideas. Perhaps the story of the weak interaction is now approaching its culmination. It remains to be seen.

The problem of the strong interaction is much more difficult than those which have faced scientists in dealing with the gravitational, electrical, and weak interactions. The mathematics appears to be more obscure, and the necessary experiments are still harder to perform. This does not necessarily mean that the physicist today is in a more difficult position than his predecessor, because the physicist today has knowledge and techniques that were not available to his predecessors, which tends to balance up the position.*

*There is one important respect in which the present-day physicist is worse off than his predecessors. His equipment costs vastly more than the equipment used fifty, or even twenty, years ago. Raising money for new equipment has thus become a major problem in physics.

But it does mean that there is little point in trying to describe here the many thousands of scientific papers which have been written on the strong interaction. The average traveler expects to drive his car along a finished highway, not to be subject to the potholes, slides, and other hazards of a road under construction. The strong interaction is emphatically a road still under construction.

The strong interaction binds the protons and neutrons of the nuclei of atoms. It may be wondered how physicists, not fully understanding the nature of this binding, are nevertheless able to produce practical nuclear devices, for example, a nuclear power plant. There is nothing new in this situation. Imperfect theories have always been able to produce important practical devices. What happens is that imperfect knowledge can often be augmented by empirical experience. The telescope, microscope, and other optical devices were invented long before the nature of light was properly understood. The electric motor and the dynamo were invented before Maxwell. Not all practical applications can precede understanding, however. Modern solid-state electronics could not have preceded quantum mechanics. Similarly, an understanding of the strong interaction might well lead to practical applications of which we cannot even conceive today. But this is not the aim of the mathematician or the physicist. Practical applications come as they may. The aim is to bite good and hard on the fruit of the tree of knowledge.

FURTHER READINGS

The classic papers on $E = mc^2$ were H. A. Lorentz, *Amsterdam Proceedings,* VI (1904), 809 and H. Minkowski, *Gött. Nach.,* 1908, 53.

For an account of the life of Maxwell, see J. R. Newman, "James Clerk Maxwell," *Scientific American,* **192** (June 1955), 58.

For a discussion of parity violation, see P. Morrison, "The Overthrow of Parity," *Scientific American,* **196** (April 1957), 45.

4

THE ASTROPHYSICIST'S UNIVERSE

ALBERT EINSTEIN
(1879–1955)

4

THE ASTROPHYSICIST'S UNIVERSE

The physicist is mainly interested in the detailed structure of the threads in our tapestry. He is less interested in the broad pattern on the tapestry itself. The broadest pattern of all, on the scale of stars and galaxies, is the business of the astronomer. The scale of our daily lives, both in spatial dimensions and in time, is already much larger than the scale in which the physicist is usually interested. Measured logarithmically,* the human body is already more than half-way from an atom to a star, and our human lifetimes are already more than half-way from the time intervals of importance in physics to the age of the Universe.

Just as we can show places on the surface of the Earth in the form of maps, so maps of the stars in the sky can be made, as in Figures 4.1 to 4.6. Different areas of the sky are described by names, just as countries on the Earth have different names. The various regions of the sky are called constellations, a complete list of which is given in Table 4.1.

*On a logarithmic scale a factor 10 counts 1, a factor 100 counts 2, 1,000 counts 3, and so on. Going from the scale of an atom to the scale of a man, the count is about 10; from a man to a star, the count is about 9.

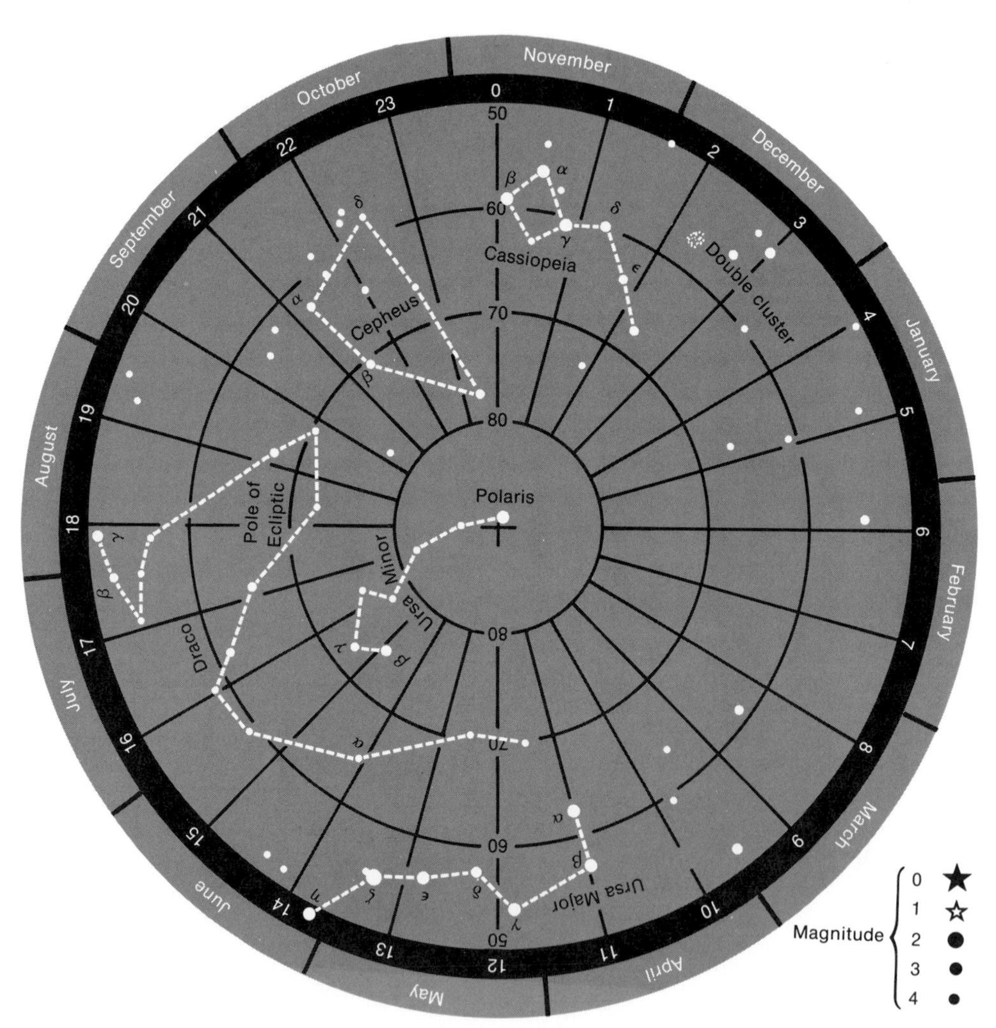

FIGURE 4.1
The constellations of the northern polar cap.

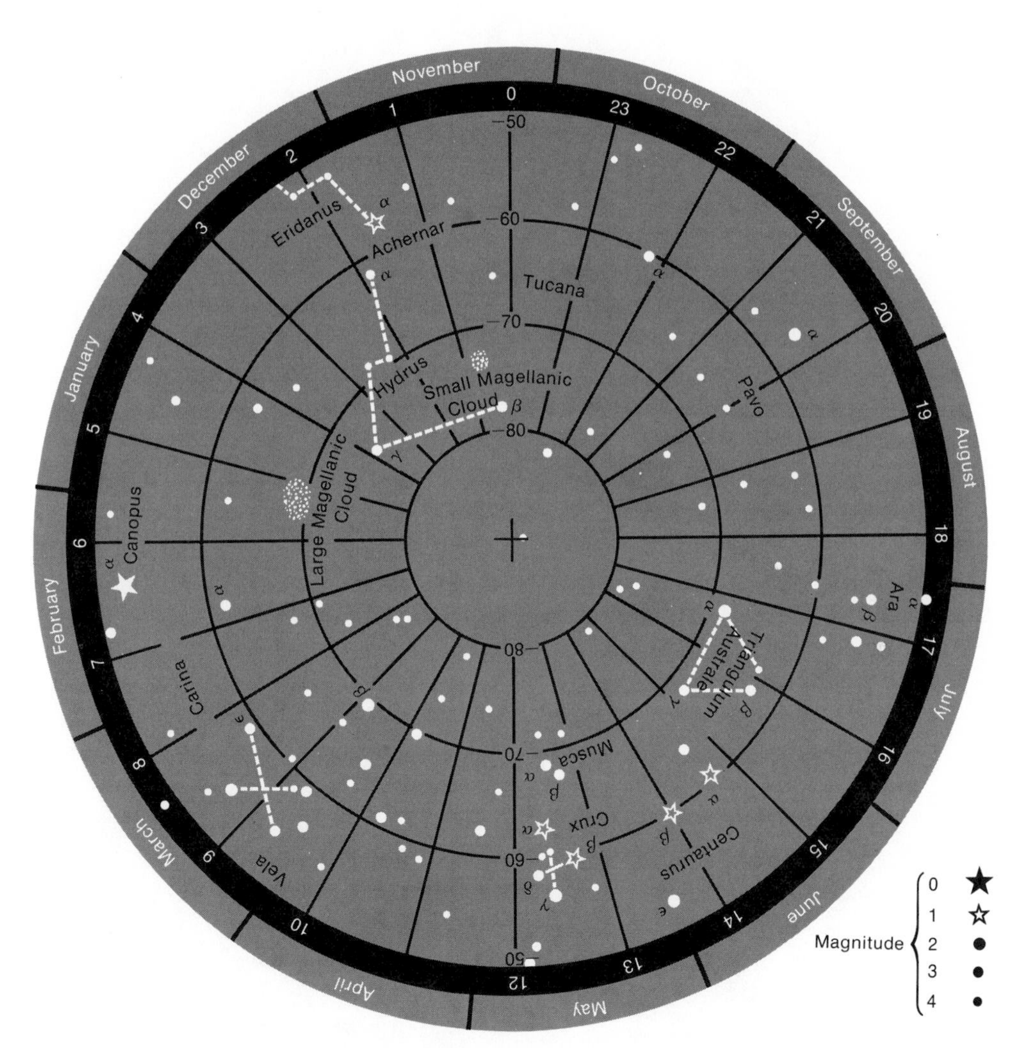

FIGURE 4.2
The constellations of the southern polar cap.

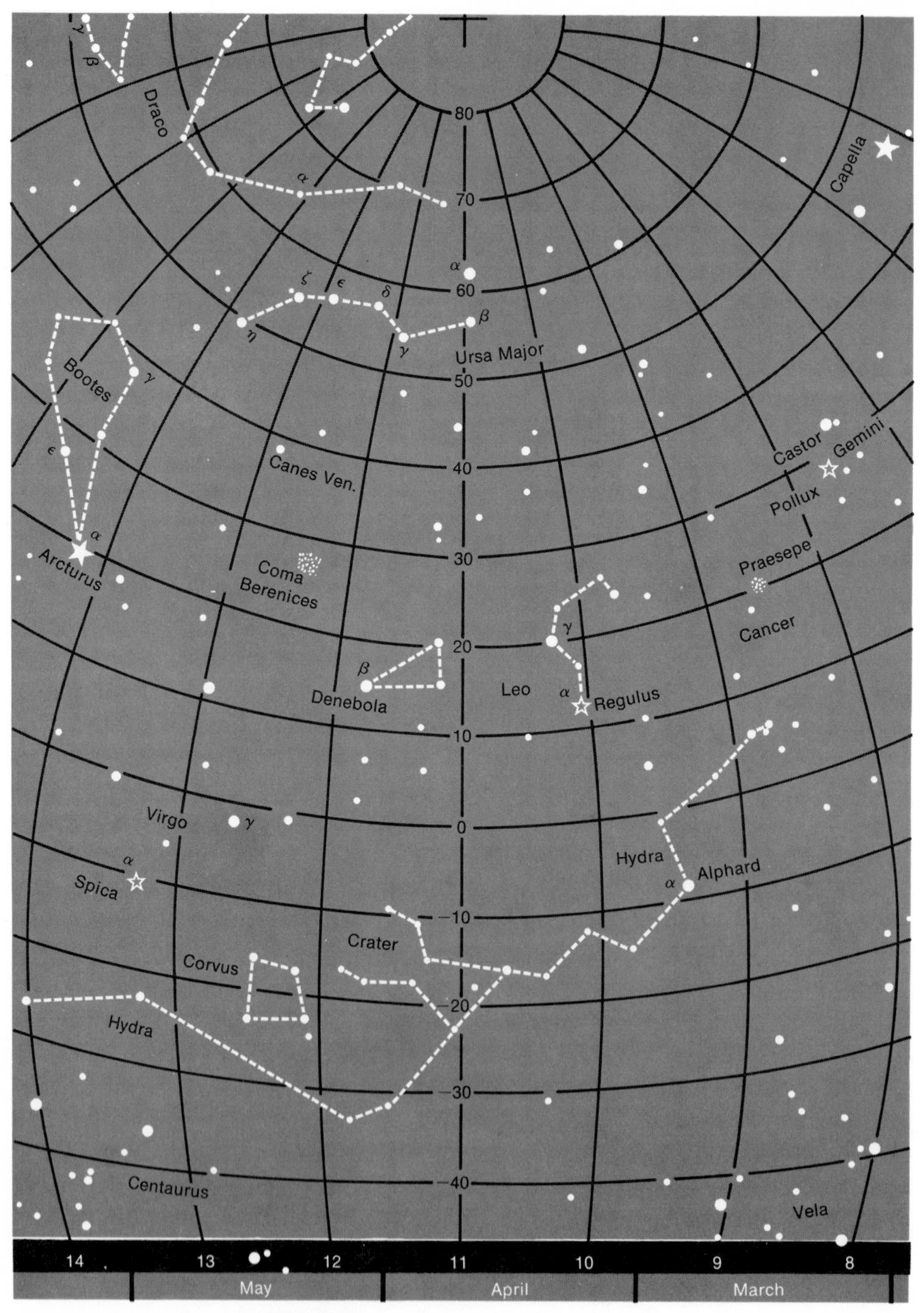

FIGURE 4.3
Night sky in Spring (northern hemisphere).

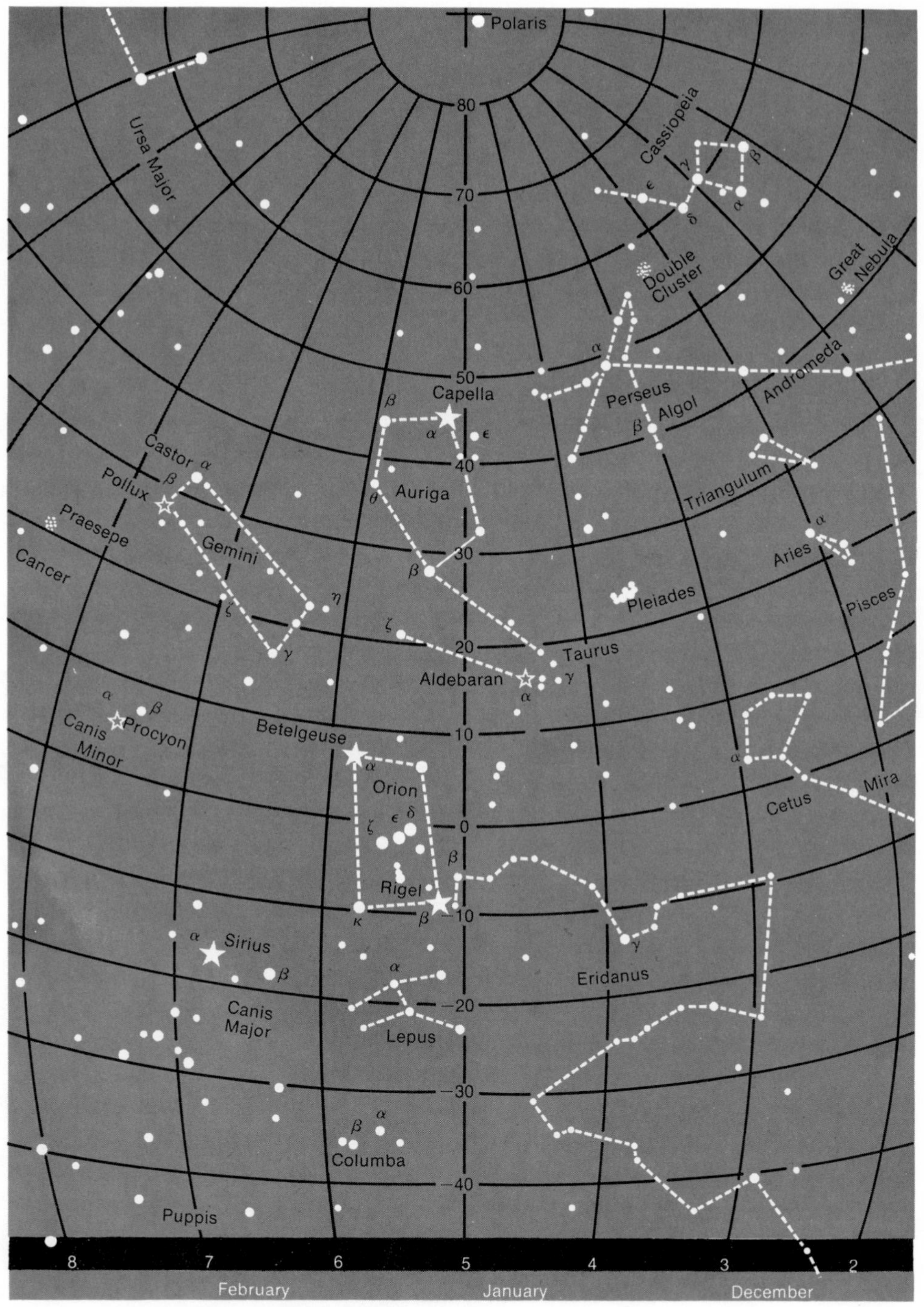

FIGURE 4.4
Night sky in Winter (northern hemisphere).

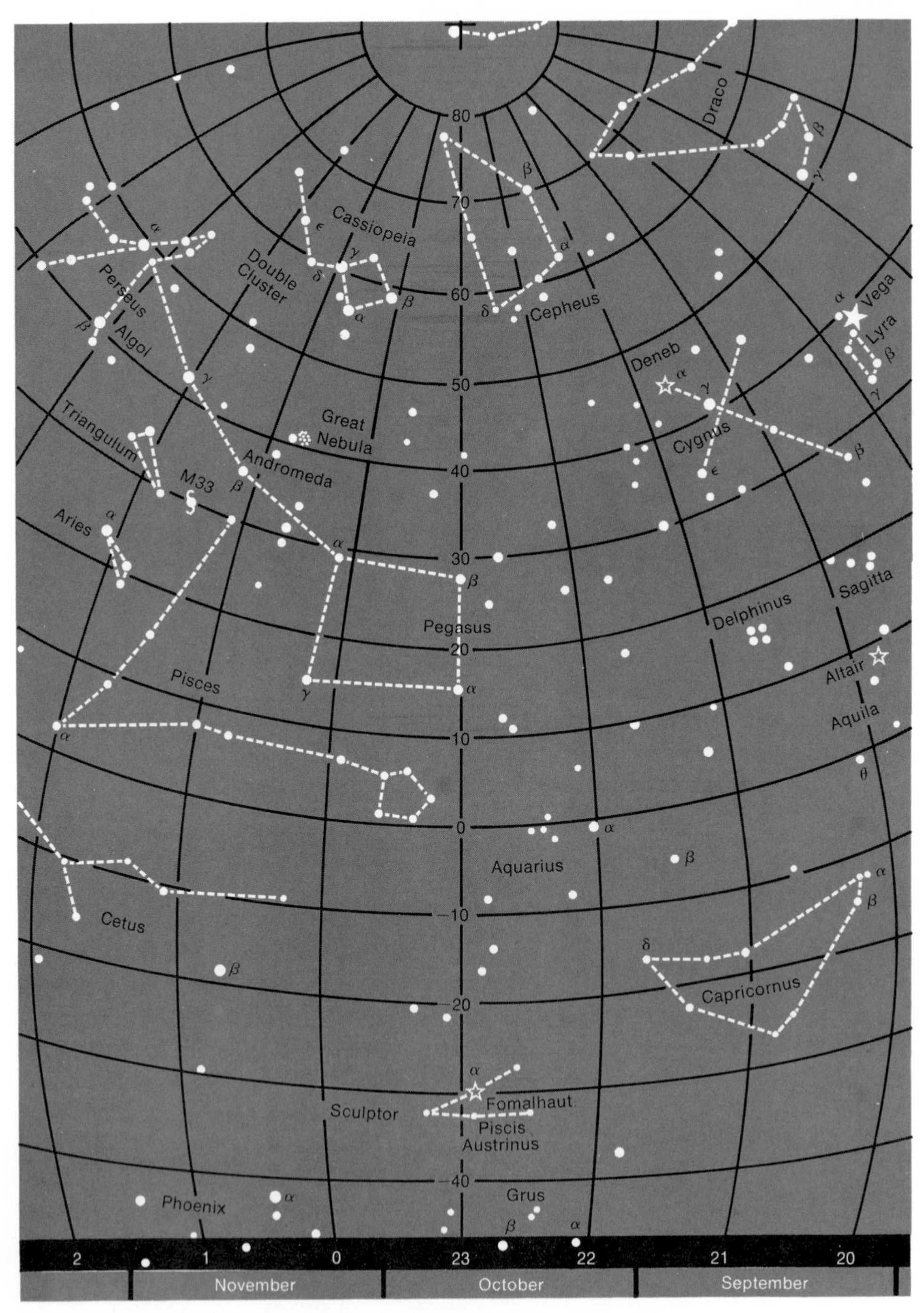

FIGURE 4.5
Night sky in Autumn (northern hemisphere).

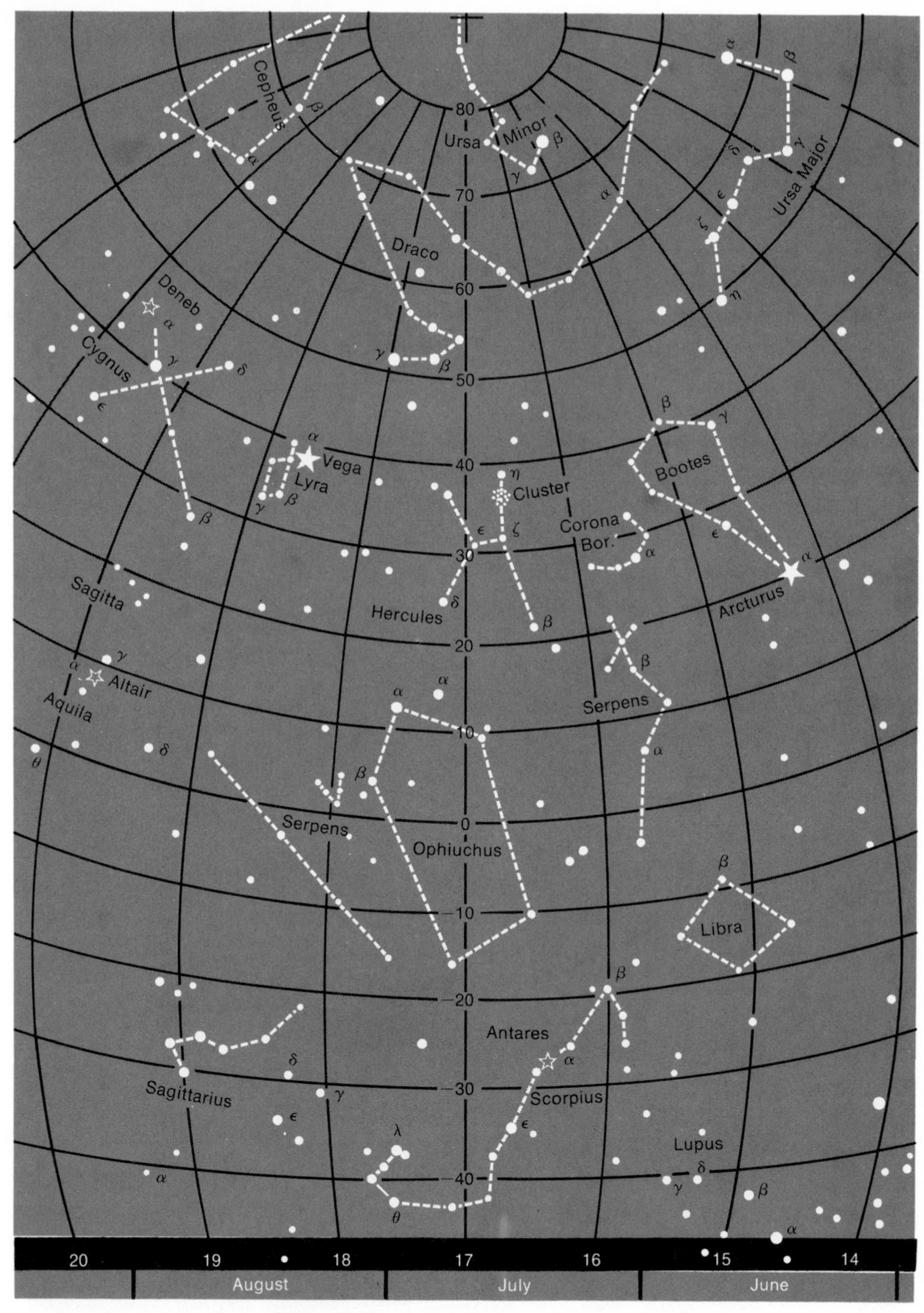

FIGURE 4.6
Night sky in Summer (northern hemisphere).

TABLE 4.1
The constellations

Name	*Approximate right ascension*	*Approximate declination*	*Intended meaning*
Andromeda (And)	01	35	Andromeda
*Antlia (Ant)	10	−30	Pump
*Apus (Aps)	17	−75	Bird of Paradise
Aquarius (Aqr)	22	−15	Water Bearer
Aquila (Aql)	20	05	Eagle
Ara (Ara)	17	−55	Altar
Aries (Ari)	02	20	Ram
Auriga (Aur)	05	40	Charioteer
Boötes (Boo)	15	30	Herdsman
*Caelum (Cae)	05	−40	Chisel
*Camelopardus (Cam)	06	70	Giraffe
Cancer (Cnc)	09	20	Crab
*Canes Venatici (CVn)	13	40	Hunting Dogs
Canis Major (CMa)	07	−25	Big Dog
Canis Minor (CMi)	07	05	Small Dog
Capricornus (Cap)	21	−15	Sea Goat
*Carina (Car)	09	−60	Ship's Keel
Cassiopeia (Cas)	01	60	Cassiopeia
Centaurus (Cen)	13	−50	Centaur
Cepheus (Cep)	21	65	Cepheus
Cetus (Cet)	02	−5	Whale
*Chamaeleon (Cha)	11	−80	Chameleon
*Circinus (Cir)	16	−65	Compass
*Columba (Col)	06	−35	Dove
*Coma Berenices (Com)	13	20	Berenice's Hair
Corona Austrina (CrA)	19	−40	Southern Crown
Corona Borealis (CrB)	16	30	Northern Crown
Corvus (Crv)	12	−20	Crow
Crater (Crt)	11	−15	Cup
*Crux (Cru)	12	−60	Southern Cross
Cygnus (Cyg)	21	40	Swan
Delphinus (Del)	21	15	Dolphin
*Dorado (Dor)	05	−60	Swordfish
Draco (Dra)	18	60	Dragon
Equuleus (Equ)	21	10	Small Horse
Eridanus (Eri)	03	−25	River Eridanus
*Fornax (For)	03	−30	Furnace
Gemini (Gem)	07	25	Twins
*Grus (Gru)	22	−45	Crane
Hercules (Her)	17	30	Hercules
*Horologium (Hor)	03	−55	Clock
Hydra (Hya)	10	−15	Water Monster
*Hydrus (Hyi)	01	−70	Water Snake
*Indus (Ind)	20	−50	Indian

TABLE 4.1 (*continued*)

Name	Approximate right ascension	Approximate declination	Intended meaning
*Lacerta (Lac)	22	40	Lizard
Leo (Leo)	10	20	Lion
*Leo Minor (LMi)	10	35	Small Lion
Lepus (Lep)	05	−20	Hare
Libra (Lib)	15	−15	Balance
Lupus (Lup)	15	−45	Wolf
*Lynx (Lyn)	09	40	Lynx
Lyra (Lyr)	19	35	Harp
*Mensa (Men)	06	−75	Table (Mountain)
*Microscopium (Mic)	21	−35	Microscope
*Monoceros (Mon)	07	00	Unicorn
*Musca (Mus)	13	−70	Fly
*Norma (Nor)	16	−55	Square
*Octans (Oct)	22	−85	Octant
Ophiuchus (Oph)	17	0	Snake Bearer
Orion (Ori)	05	00	Orion
*Pavo (Pav)	20	−60	Peacock
Pegasus (Peg)	22	20	Pegasus
Perseus (Per)	03	40	Perseus
*Phoenix (Phe)	01	−45	Phoenix
*Pictor (Pic)	07	−60	Easel
Pisces (Psc)	00	10	Fishes
Piscis Austrinus (PsA)	23	−30	Southern Fish
*Puppis (Pup)	07	−35	Ship's Stern
*Pyxis (Pyx)	09	−35	Ship's Compass
*Reticulum (Ret)	04	−65	Net
Sagitta (Sge)	20	15	Arrow
Sagittarius (Sgr)	18	−30	Archer
Scorpius (Sco)	17	−35	Scorpion
*Sculptor (Scl)	01	−30	Sculptor
*Scutum (Sct)	19	−10	Shield
Serpens (Ser)	16	05	Snake
*Sextans (Sex)	10	00	Sextant
Taurus (Tau)	05	20	Bull
*Telescopium (Tel)	18	−45	Telescope
Triangulum (Tri)	02	35	Triangle
*Triangulum Australe (TrA)	16	−65	Southern Triangle
*Tucana (Tuc)	23	−60	Toucan
Ursa Major (UMa)	11	50	Great Bear
Ursa Minor (UMi)	15	75	Small Bear
*Vela (Vel)	09	−50	Ship's Sails
Virgo (Vir)	13	00	Virgin
*Volans (Vol)	08	−70	Flying Fish
*Vulpecula (Vul)	20	25	Fox

*Of modern origin.

The names of the constellations developed from association of star patterns with quaint terrestrial images. These names have no scientific significance, but they continue to have aesthetic attractions for us all. The numbers given in Table 4.1 represent an astronomical system of latitude (declination) and longitude (right ascension). These numbers can be used to locate the different constellations in the star maps of Figures 4.1 to 4.6.

It is a curious thought that modern man is probably less familiar with the night sky than were our remote ancestors of 100,000 years ago. Life in cities tends to change our mode of existence to a state in which we are hardly aware of the sky or, indeed, of many other aspects of our natural environment.

The astrophysicist is interested (as his name implies) both in the stars themselves and in the atoms which make up the stars—he is interested in both cables and threads. There are many types of atom, as we saw in Chapter 2. To recapitulate what was said there, an atom is determined by the neutrons and protons within its nucleus. Two atoms having the same number of protons but with different numbers of neutrons are given the same name. They are said to belong to the same element, being isotopes of that element. The element copper has two isotopes, for example, one with 29 protons and 34 neutrons, the other with 29 protons and 36 neutrons. The element is different, however, when two atoms have different proton numbers. Denoting the proton number of an atom by Z, the value of Z determines the element, as it was shown to do in Table 2.1.

The righthand column of Table 2.1 gives the relative abundances of the elements as they are present in "cosmic material." The Sun is made up of the various kinds of atom in these proportions, and since most of the stars of Figures 4.1 to 4.6 are rather similar to the Sun, they too have compositions similar to that given in Table 2.1. It is just because this abundance distribution is very widespread on an astronomical scale that the term "cosmic material" is used to describe these relative proportions of the elements. Notice the reference standard of exactly 1 million for the element silicon. The other abundances are all stated relative to this particular standard.

The Earth is not composed of cosmic material. On the Earth we have a *selection* of the elements, with a few of the cosmically very abundant elements, like helium, present only in much reduced concentrations. Of the other planets of our system, helium is also largely absent from Mercury, Venus, Mars, Uranus, and Neptune, but is present on the planets Jupiter and Saturn, where the helium abundance is like that of the Sun. In Chapter 8 we shall see that differences of this kind provide clues to the way in which our whole system of planets was originally formed.

The selection process in which the Earth was formed did not operate, however, to make low abundances into high abundances. Elements of large Z (greater than about 30) have low abundances both cosmically and on the Earth.

The economic value of various metals tends in a general way to follow the cosmic abundance values. The commonest and least expensive metals, magnesium, aluminum, iron, have rather high abundances. Zinc and copper, being less abundant, are more valuable. Tin is still less abundant and still more valuable. The precious metals, silver and gold, platinum, palladium, rhodium, and iridium, are all cosmically rare elements. Yet the relation between economic value and cosmic abundance is not entirely strict. Naturally, economic value depends on usefulness, and also on the way in which the various metallic ores were deposited in the crustal rocks of the Earth. Some elements tend to concentrate into comparatively rich deposits; others remain more diffusely distributed and are therefore harder to obtain. The element bismuth is cosmically about as rare as gold; yet the economic value of bismuth is much less than that of gold, since rich deposits of bismuth are accessible, whereas rich deposits of gold are not. Provided we keep these further considerations in mind, the cosmic abundances of Table 2.1 provide us with a useful general guide to the economic values of the elements.

It is interesting and important that life itself is based on the commonest elements. Of all substances, water is perhaps the most essential to life, and the molecule of water, H_2O, is composed of the cosmically most abundant element (H_2) in combination with the third most abundant element (O). Carbon and nitrogen form the backbone of living material, and these are the fourth and fifth commonest elements, respectively. Other important elements, calcium in our bones, iron in the blood, magnesium in the chlorophyll of plants, sulfur, and phosphorus, are all elements of rather high abundance, much higher than the elements of large Z. It is the rarity of the latter which prevents them from playing an important part in the structure of life. Life is built from the simplest molecules containing the commonest elements, suggesting that life may well be cosmically very abundant indeed.

The distribution of the atoms and of their isotopes is totally populated, meaning that every possible combination of protons and neutrons into the nuclei of atoms is actually found on the Earth. A possible combination is one that can persist for as long as the age of the Earth (about 4,600 million years) without a proton changing to a neutron, or a neutron changing to a proton—i.e., without the atom itself being altered to a different atom. Does the fact that we can find every possible combination of protons and neutrons on Earth mean the atoms have always existed, or have they come into being in some way?

A similar question faced biologists early in the nineteenth century. Classifications of plants and animals had by then been made, revealing an astonishing range of species and of varieties of species. Had all this richness appeared in one brief flash of creation, or had some continuous evolutionary process operated to build the observed life forms, a process using only the simplest chemical substances—water, nitrogen, carbon? The enormous complexity of what would be needed for such an evolution overwhelmed the imagination of most biologists of the period, who therefore sought refuge in the apparently less difficult idea of a special moment of creation. But Nature is not subject to the immediate limitations of the human imagination, and the concept of evolution, which once seemed incredible, is nowadays seen to be correct.

The cosmic abundances of the elements indicate that matter has also evolved from simpler forms to more complex ones. For matter, hydrogen atoms are the starting point in an evolutionary process whereby more complex atoms are formed. The process occurs inside stars. Hydrogen goes to helium inside the stars marked in Figures 4.1 to 4.6, helium goes to carbon and oxygen within at least some of these same stars; and carbon and oxygen go to more complex atoms within certain remarkable stars that are detectable instrumentally, but are not normally seen by eye, and so are not shown in Figures 4.1 to 4.6. Since stars play a critical role in the history of matter, and so set the stage for the origin of life, it is of interest to consider their birth, their lives, and their death.

Stars are being born right now within clouds like the Orion Nebula, shown in Figure 4.7. Using binoculars, one can easily see this glowing mass of gas in the "sword" of the constellation of Orion. The diameter of the Orion Nebula is about 15 light years, a light year being simply the distance traveled by light in a year, about 10 million million kilometers. The diameter of the cloud shown in Figure 4.7 is therefore about 150 million million kilometers. There is enough gas in it to form a very large number of stars, perhaps as many as 100,000 stars.

The distance of the Orion Nebula from us is about a hundred times greater than its diameter, about 1,500 light years, and the over-all diameter of the great star system in which we live, often referred to colloquially as the Milky Way, is nearly a hundred times greater still, about 100,000 light years. There are many gas clouds to

FIGURE 4.7
The Orion Nebula, a cloud of gas in which stars are now being born. (Courtesy of the Hale Observatories.)

FIGURE 4.8
The Rosette Nebula. (Courtesy of the Hale Observatories.)

be found along the Milky Way. The Orion Nebula happens to be one of the nearest. Another example, the Rosette Nebula, is shown in Figure 4.8. Stars appear to be forming now in all these gas clouds.

Stars pour out energy from their surfaces. We are well used to the flood of light and heat which the Earth receives from the Sun. How, we may ask, does the Sun manage to go on providing power at the enormous rate of 3.8×10^{23} kilowatts, and how has the energy loss been made good during the whole time span of the Sun's past history (nearly 5,000 million years)?

The gas deep inside the Sun is hot, the temperature rising at the solar center to about 15 million degrees. When the protons, which constitute the nuclei of hydrogen atoms, collide with each other at this temperature, the weak interaction (Chapter 3) gradually changes some of them to neutrons. What happens is that an "up" quark within the triplet which constitutes a proton changes to a "down" quark. The neutron so formed then joins with another proton to form a nucleus with one neutron and one proton, a nucleus known as the deuteron. Then by further steps, again due to collisions at the high temperature, the deuterons join in pairs to form the nuclei of helium atoms. The net effect is to change hydrogen into helium. Energy is released in this sequence of reactions, and it is the energy so generated which makes good the radiation loss from the surface of the Sun. The source of the energy which the Earth receives from the Sun is therefore *nuclear*. It has been obtained by transmuting atoms of hydrogen into atoms of helium, a transmutation of a kind that was avidly sought by the medieval alchemists, and which is being sought just as avidly by modern experimental scientists. If hydrogen could be converted to helium under controlled industrial conditions, the "burning" of only 3,000 tons of water would be sufficient to operate the world's industries for a whole year.

The large mass of the Sun, about 300,000 times the mass of the Earth, taken with the high cosmic abundance of hydrogen, explains the apparently prodigious amount of energy which the Sun has generated throughout its history. The hydrogen still remaining inside the Sun is sufficient to maintain the present power output of 3.8×10^{23} kilowatts for a further 10,000 million years. A vast future for life on the Earth is therefore possible, provided Man proves capable of seizing the opportunity to sustain and develop this particular oasis of life in the universe.

Stars are by no means alike. They differ in mass, and they differ especially in the rate at which they emit light and heat. A star with a mass ten times that of the Sun emits light and heat at more than a thousand times the rate of the Sun. To compensate for this prodigal energy emission, such a more massive star must convert hydrogen to helium far more rapidly than the Sun does. The rapid use of hydrogen in a star of large mass causes its hydrogen supply to become depleted in a comparatively short time. Whereas the hydrogen in the Sun will last for 10,000 million years, hydrogen in a star of large mass can last for no more than a few tens of millions of years, a time span much less than the age of the Milky Way, an age which has been estimated to be about 12,000 million years. Hydrogen in such prodigal massive stars therefore becomes exhausted on a time scale that is, cosmically speaking, rather short. What, we may wonder, happens next?

Many nuclear reactions beside those which convert hydrogen to helium are possible. What is special about the reactions that convert hydrogen to helium is that they are the first to become important as the temperature rises inside a newly formed star. The further contraction of the central region of the star, which sets in as the hydrogen becomes exhausted, raises the temperature further; and other, higher-temperature reactions come into operation. The helium itself can undergo reactions that produce the elements carbon and oxygen. This "burning" of helium is the next stage in the evolutionary process whereby simpler atoms are built into more complex ones. At higher temperatures still, carbon and oxygen themselves undergo reactions which generate a whole host of well-known elements: sodium, magnesium, aluminum, silicon, sulfur, and calcium. At even higher temperatures, the common metals appear: iron, nickel, chromium, manganese, and cobalt.

These are the main processes that give rise to the elements with comparatively large abundances. Elements of low abundance are also generated in the stars, but in more minor by-product reactions.

We see, then, how it comes about that the evolution of atoms is associated with the evolution of the stars themselves. By means of nuclear processes that take place inside stars, the familiar materials of our everyday world have been produced; the stars are intensely hot furnaces in which the everyday materials have been forged from the simplest atom, hydrogen. These same processes have also served the stars as an energy source, enabling them to emit radiation from their surfaces for long periods of time. For stars of comparatively small mass, the evolution from hydrogen to the iron-like elements is by no means yet completed. For most stars of large mass, however, there has been ample time for the whole evolutionary sequence to reach its end point. What, we may ask, is this end point?

The complex network of nuclear reactions, driving matter from hydrogen up to complex atoms like iron, provides a star with only a limited amount of energy. Each new stage of the evolutionary process, coming into operation as the inner regions of the star become hotter and hotter, provides successively less energy, and all the evolutionary stages taken together are like a limited number of checks which the star can cash. Once they are cashed, no more energy can be forthcoming from nuclear processes.

We might guess that the star must then cool off, becoming a dead star which ceases to shine. But calculation shows that a star cannot die in this way unless the mass is less than a certain upper limit, which turns out to be not too different from the mass of the Sun. In order to be able to cool off, a star more massive than this must fling material out into space—from where all its material originally came. There are two main ways in which this shedding of excess material seems to happen.

Stars of moderate mass shed most of their unwanted material rather gently, probably ending with a fling in which a cloud known as a planetary nebula is thrown off. Examples of planetary nebulae are shown in Figures 4.9 and 4.10. The residue left behind eventually cools to become a remarkable kind of star known as a *white dwarf.*

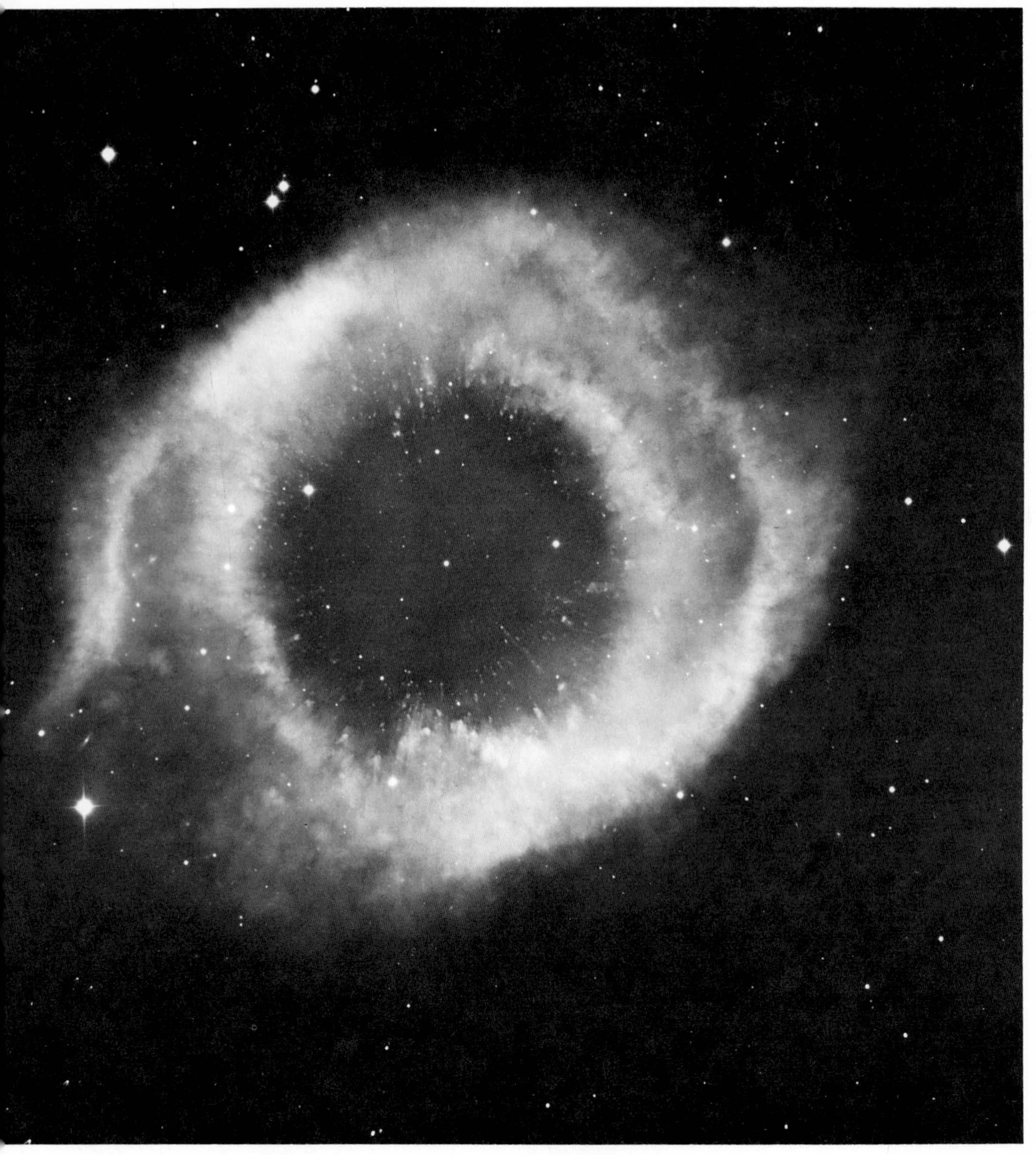

FIGURE 4.9
A planetary nebula, NGC 7293. (Courtesy of the Hale Observatories.)

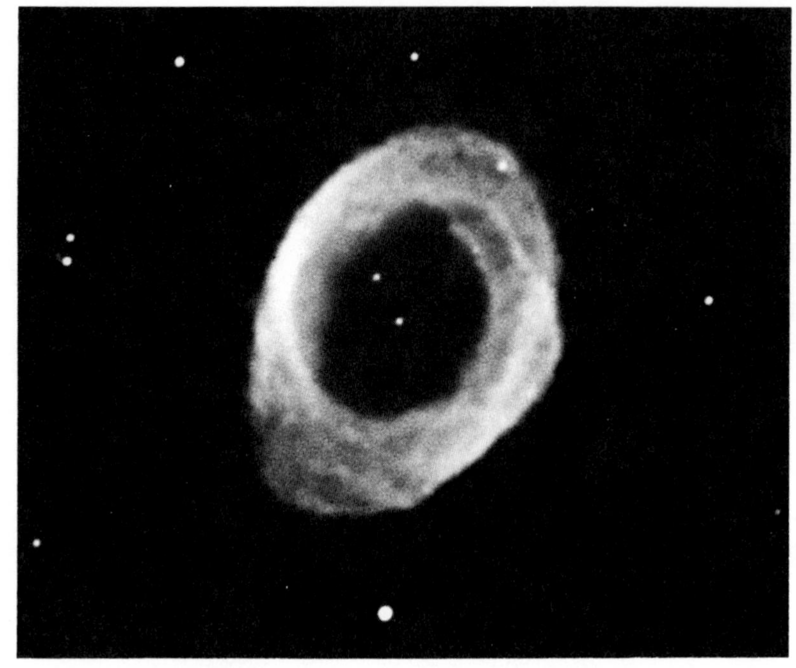

FIGURE 4.10
The Ring Nebula, NGC 6720, another planetary nebula. (Courtesy of the Hale Observatories.)

White dwarfs, although they have masses quite comparable to that of the Sun, have diameters comparable to those of planets, which implies exceedingly high densities within them. The material at the center of a white dwarf can have a density so high that a chunk the size of an ordinary sugar cube contains a ton of material.

Stars of large mass are still more violent in their behavior. They become unstable, and explode like a nuclear bomb. For a few days following such an explosion, usually referred to as a *supernova,* the star is temporarily as bright as the whole of the Milky Way. Such explosions occur in our galaxy with a frequency of about one per century. A supernova observed by Chinese astronomers in the year A.D. 1054 produced the object shown in Figure 4.11. From its appearance this object was named (long before its nature was understood) the Crab Nebula. The supernova was seen by Chinese

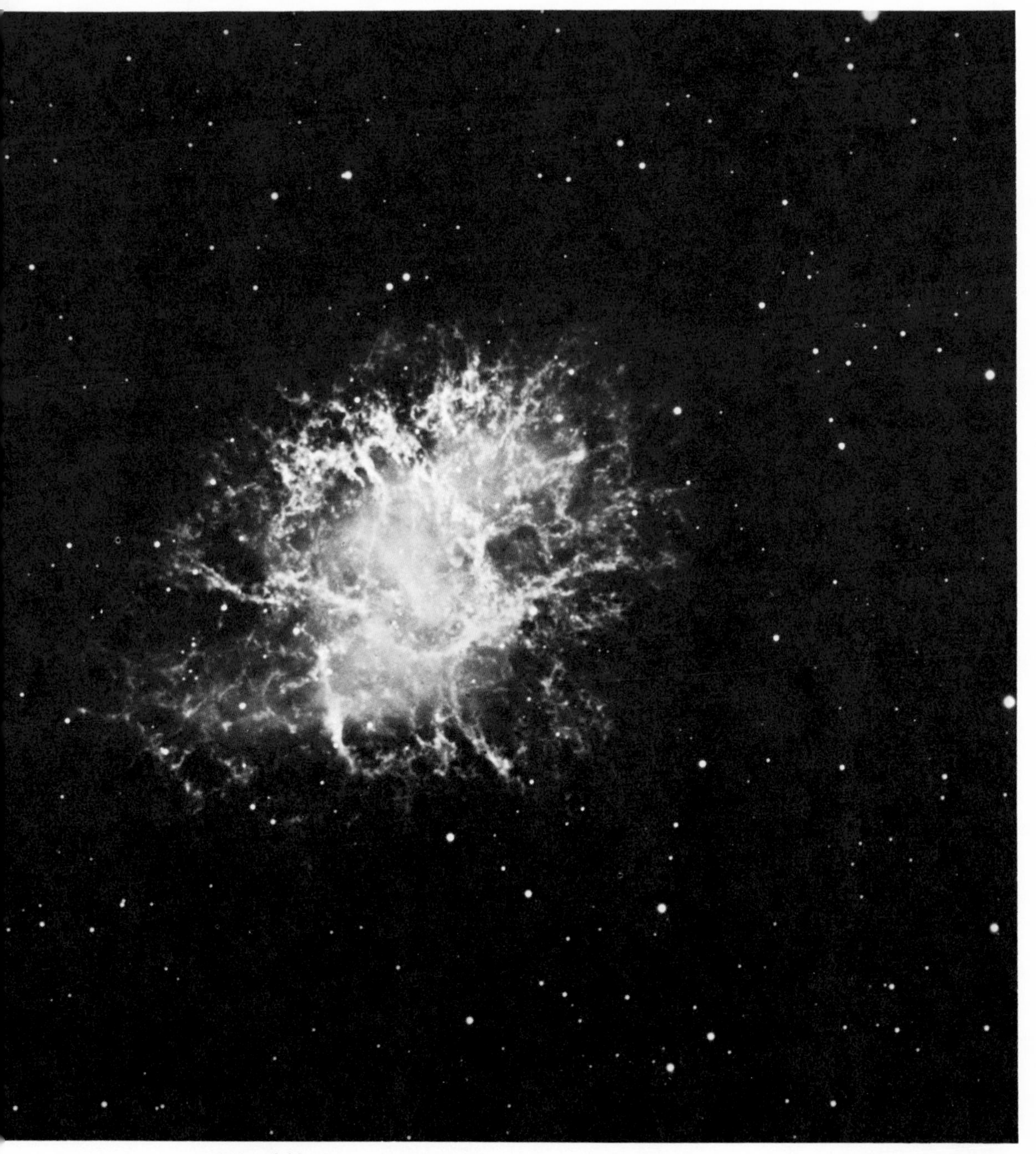

FIGURE 4.11
The Crab Nebula. The light from this object is generated by electrons of very high speed moving in a magnetic field. (Courtesy of the Hale Observatories.)

Sun

White dwarf Neutron star Earth

FIGURE 4.12
Neutron stars are much smaller than the planets.
(After Gilluly, Waters, and Woodford, *Principles of Geology,*
3d ed. W. H. Freeman and Company. Copyright © 1968.)

astronomers on July 4, 1054, as a "guest star visible by day like Venus." It remained as bright as Venus until July 27. Thereafter it faded gradually, becoming invisible to the naked eye by April 17, 1056. These violently exploding massive stars also leave a residue behind, a residue even more remarkable than a white dwarf. The core of such massive stars becomes too dense to be a white dwarf; it attains a fantastic condition in which a chunk the size of a sugar cube contains 100 million tons of material. They become objects known as *neutron stars.* Although they have masses comparable to that of the Sun itself, neutron stars are much smaller in size than the planets, as can be seen from Figure 4.12. Indeed, neutron stars are not far removed from the condition which astronomers and physicists refer to nowadays as a *black hole,* a condition where matter collapses in on itself and ultimately disappears altogether from our universe.*

Because of their small diameters, the neutron stars spin around very quickly, most of them about once a second, although exceptional ones may spin even faster. As they spin, a lighthouse effect seems to operate, in which all kinds of radiation—radiowaves as well as visible light, and sometimes even x-rays—sweep periodically across an external observer. When this happens, the star is known as a *pulsar.* There is a pulsar within the Crab Nebula. It turns very

*Indeed, there is some evidence that supernovae do sometimes produce a black hole rather than a neutron star.

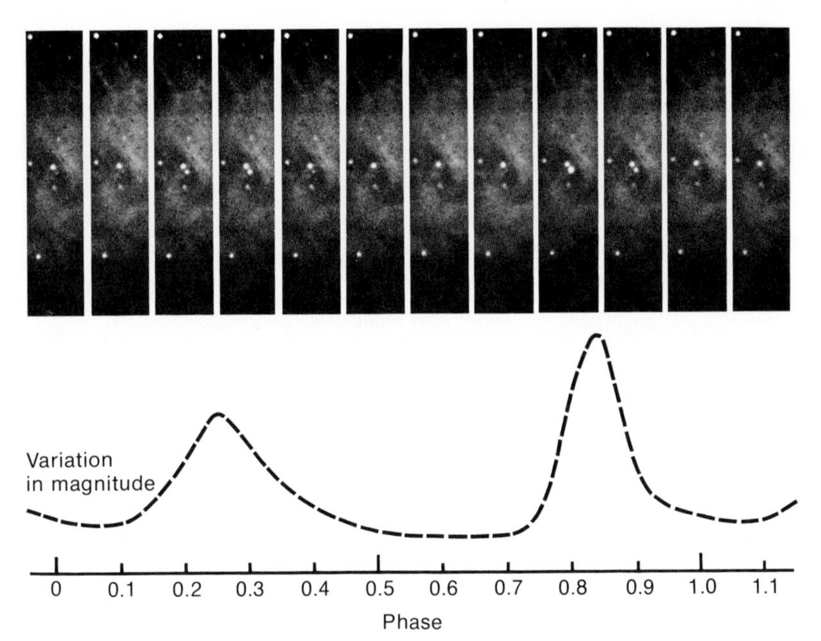

FIGURE 4.13
A variable star-like object in the Crab Nebula is believed to be a rotating neutron star. (Courtesy of the Kitt Peak National Observatory.)

rapidly, at over thirty rotations per second, and as it does the light from it goes on and off, as can be seen from the remarkable series of photographs in Figure 4.13.

By ejecting material into space at the ends of their lives, stars generate a cyclical process. They begin by condensing from a gas cloud like the Orion Nebula, and they end by returning matter to similar gas clouds. The matter which is so returned differs in its composition from the original material. The matter now contains complex atoms, atoms with many protons and neutrons in their nuclei. So the compositions of the gas clouds themselves are changed by the nuclear processes occurring within stars. Stars that

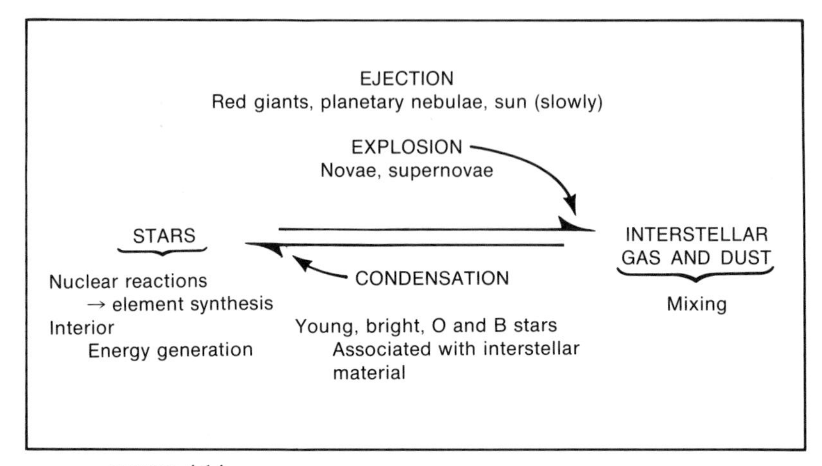

FIGURE 4.14
A cyclical process, in which matter is transferred back and forth between stars and the interstellar gas.

are forming now must therefore contain more complex atoms than did stars which formed early in the history of the Milky Way. The logical structure of this cyclical evolution, with material transferred back and forth between stars and gas clouds, where it is then available to become a part of new stars when they form, is shown in Figure 4.14.

The age of the Milky Way is at least twice the age of our planetary system. Hence by the time our system was formed, the cyclical process of Figure 4.14 had already been operating for probably 7,000 million years. The complex atoms of our local world, in the abundances given in the righthand column of Table 2.1, were produced during this long time-span. The carbon and nitrogen in our bodies, the oxygen we breathe, the iron in our blood, were all generated inside stellar furnaces at remote epochs in the past. The parent stars of our complex materials are by now dead white dwarfs or neutron stars, which we have no means of tracing and identifying. From the standpoint of biology, our presence on the Earth depends on a remarkable and even fantastic sequence of chemical processes. From the standpoint of physics, the very material of which we are constituted has experienced an evolution scarcely less remarkable.

FURTHER READINGS

W. A. Fowler, *Nuclear Astrophysics.* American Philosophical Society, 1965.

Frontiers in Astronomy: Readings from Scientific American. W. H. Freeman and Company, 1970.

5

THE EXPANDING UNIVERSE

EDWIN POWELL HUBBLE
(1889–1953)

(Photo by J. R. Eyerman,
Time-Life Picture Agency.)

5

THE EXPANDING UNIVERSE

This chapter is concerned with galaxies. The study of their motions, which we shall come to in the second half of the chapter, leads to the concept of the expansion of the universe. The investigation starts on firm ground, but it ends in shifting sands. The strange situation reached at the end will be considered further in Chapter 6, where the problem of the "origin" of the universe, perhaps the most crucial problem facing astronomy today, will be discussed.

A galaxy is a collection of stars and gas clouds like the Milky Way. Indeed, the Milky Way is often referred to as "our" galaxy. A large galaxy like our own typically contains some 10^{11} to 10^{12} stars.

The nearest other large galaxy, shown in Figure 5.1, is about 2 million light years distant from us. The central part of this external galaxy is easily seen with the naked eye. It lies in the constellation of Andromeda, and is marked "Great Nebula" in the star map of Figure 4.5. In shape both our galaxy and the one in Andromeda are flattened circular disks, each with a moderate-sized bulge developing

FIGURE 5.1
The galaxy M31, often referred to as the Andromeda Galaxy.
(Courtesy of the Hale Observatories.)

toward the center. The Andromeda galaxy appears to be an oval in Figure 5.1 because we are looking at it sideways; one can judge its orientation by drawing a circle on a piece of paper and then turning the paper to produce a similar oval shape. The stars which can be seen scattered over Figure 5.1 are simply foreground stars of our own galaxy—we have to look through our own galaxy to see the systems which lie outside.

Although the distance from our galaxy to the Andromeda galaxy is enormous when expressed in a conventional way, about 20 million million million kilometers, the distance becomes quite small when the scale of the galaxies themselves is used as a yardstick. If we choose the distance scale so that our galaxy has a diameter of just 1 unit, which we call the "yard," the average spacing between galaxies is about 100 yards. Galaxies exist cheek-by-jowl compared to the stars within a galaxy. In terms of their own scales, stars are widely spaced, whereas galaxies are rather closely spaced.

The stars of our own galaxy thin out as the confines of the Milky Way are reached. A similar thinning out on a universal scale of the galaxies themselves has been looked for by astronomers, but has not been found. The distribution of galaxies, 10^8 to 10^9 of them, goes on and on to the limits of detection by the largest telescopes.

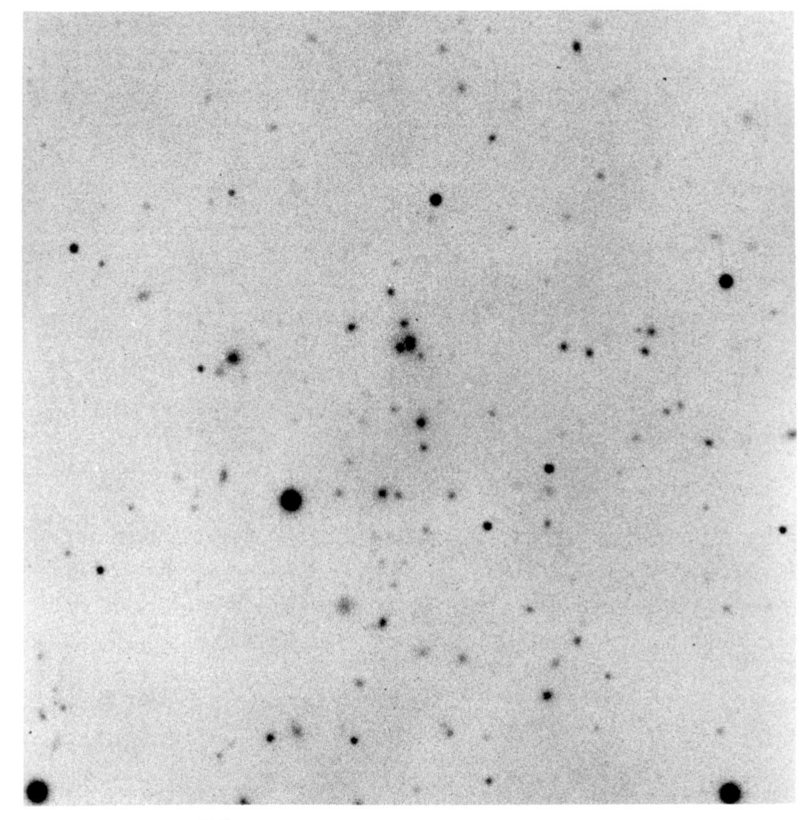

FIGURE 5.2
In the depths of space, very many galaxies are faintly seen. This cluster is in the constellation of Hydra. (Courtesy of the Hale Observatories.)

How galaxies appear when photographed near the limit of a large telescope is shown in Figure 5.2.

Examination of photographs of individual galaxies shows that they are by no means alike (although our galaxy is thought to be quite similar to the one in Andromeda, Figure 5.1). Some have remarkable spiral patterns, as in Figure 5.3; others are smooth in appearance with elliptical outlines, as in Figure 5.4.

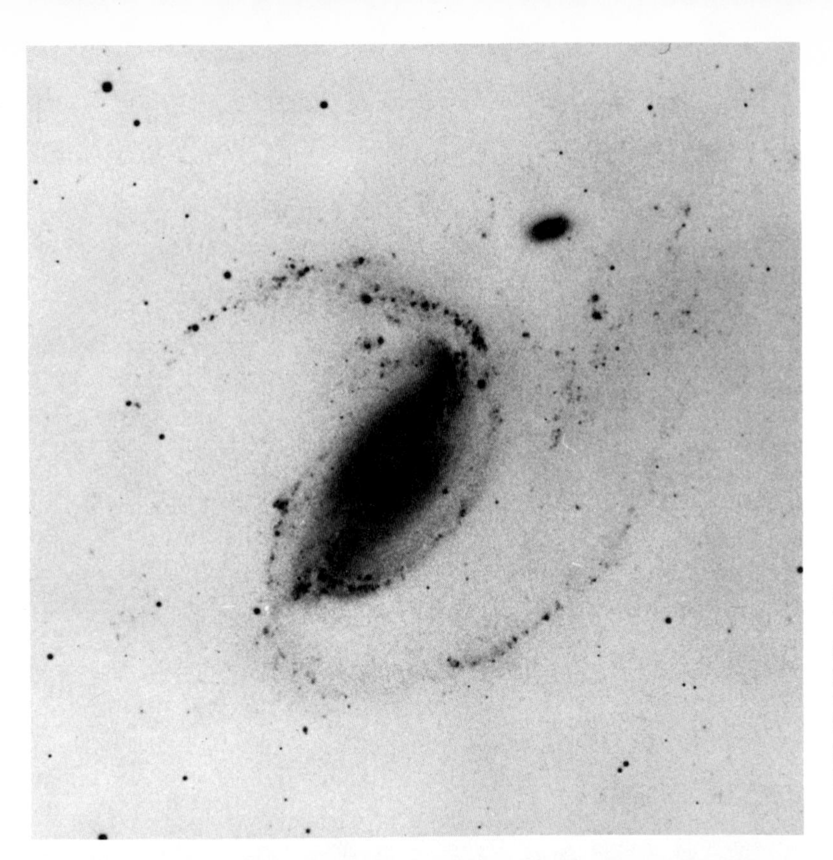

FIGURE 5.3
The remarkable spiral galaxy NGC 1097, photographed in negative form in the light emitted by hydrogen atoms. (Courtesy of Dr. H. C. Arp.)

FIGURE 5.4
The elliptical galaxy M87. (Courtesy of the Hale Observatories.)

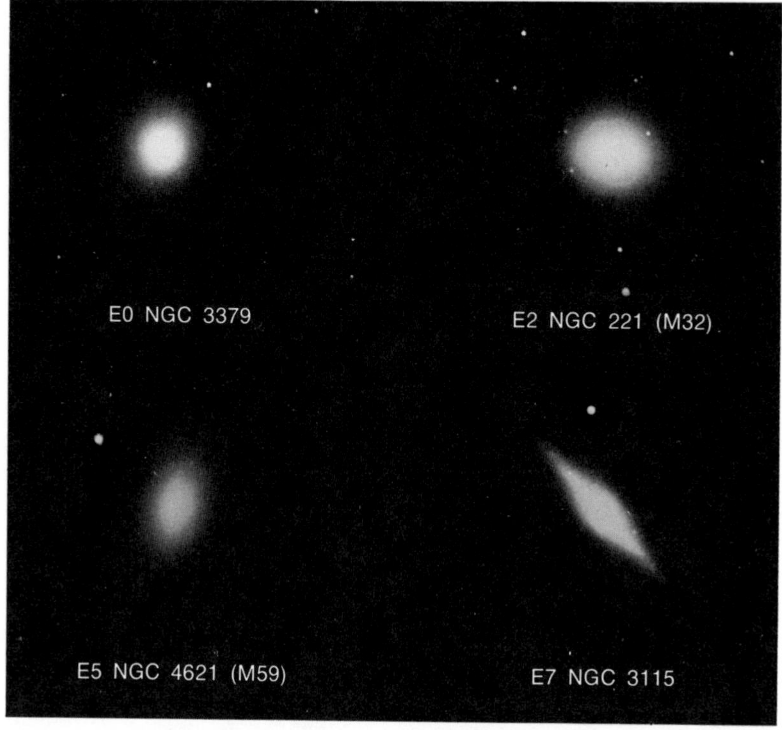

FIGURE 5.5
Examples of Hubble's elliptical sequence E0 to E7. (Courtesy of the Hale Observatories.)

The rich variety of galactic forms is illustrated by Figures 5.5 to 5.7, the designations given in these figures being those devised by Edwin Hubble (1889–1953). In this form of classification our galaxy and the Andromeda galaxy are of type Sb. The apparently rather rare galaxies falling outside this scheme, like those shown in Figures 5.8 and 5.9, were regarded by Hubble as "irregular." Recently astronomers have become very interested in these irregular galaxies, and it is now thought the peculiarities which they exhibit are not quite as rare as Hubble supposed. Even apparently normal galaxies when

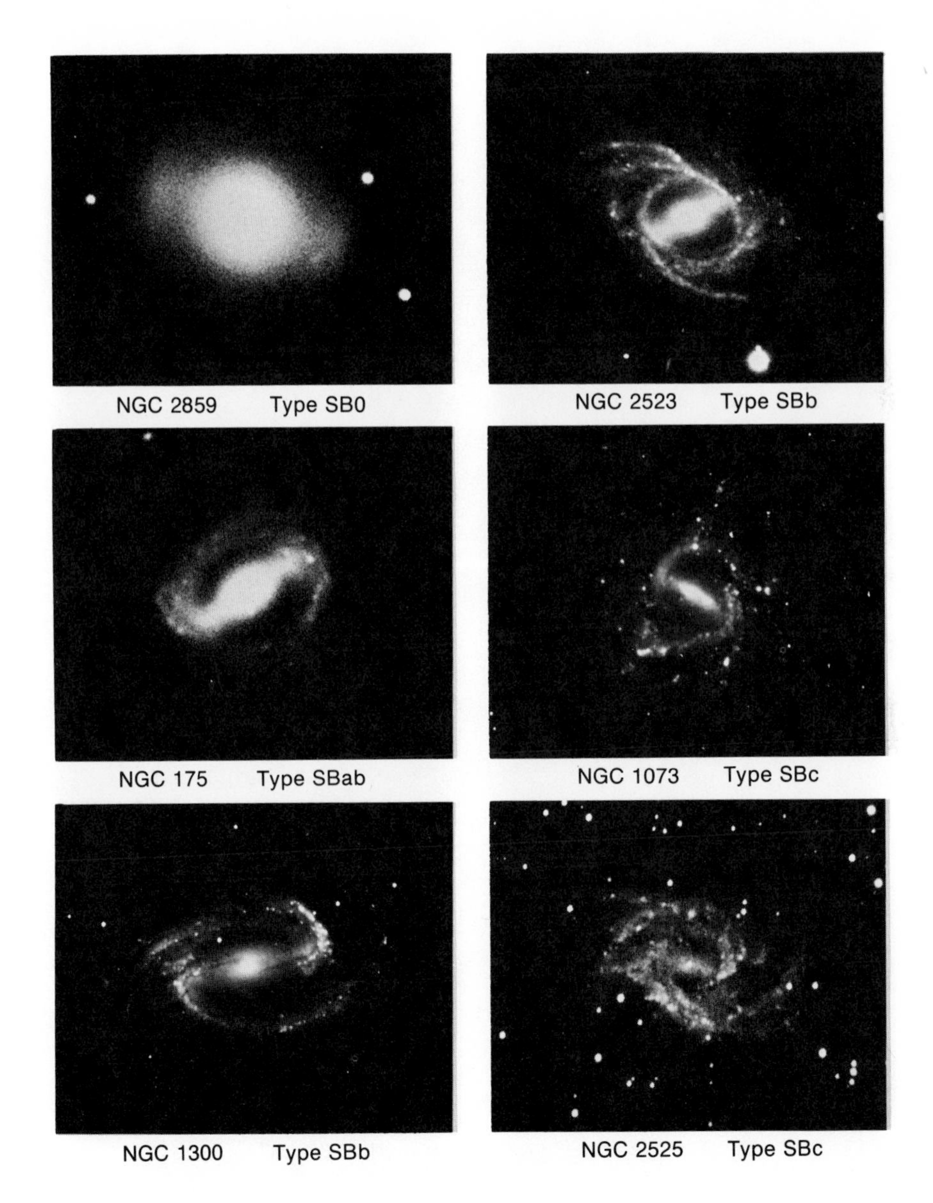

FIGURE 5.6
Examples of Hubble's Sa, Sb, Sc types, with an intermediate Sab form. (Courtesy of the Hale Observatories.)

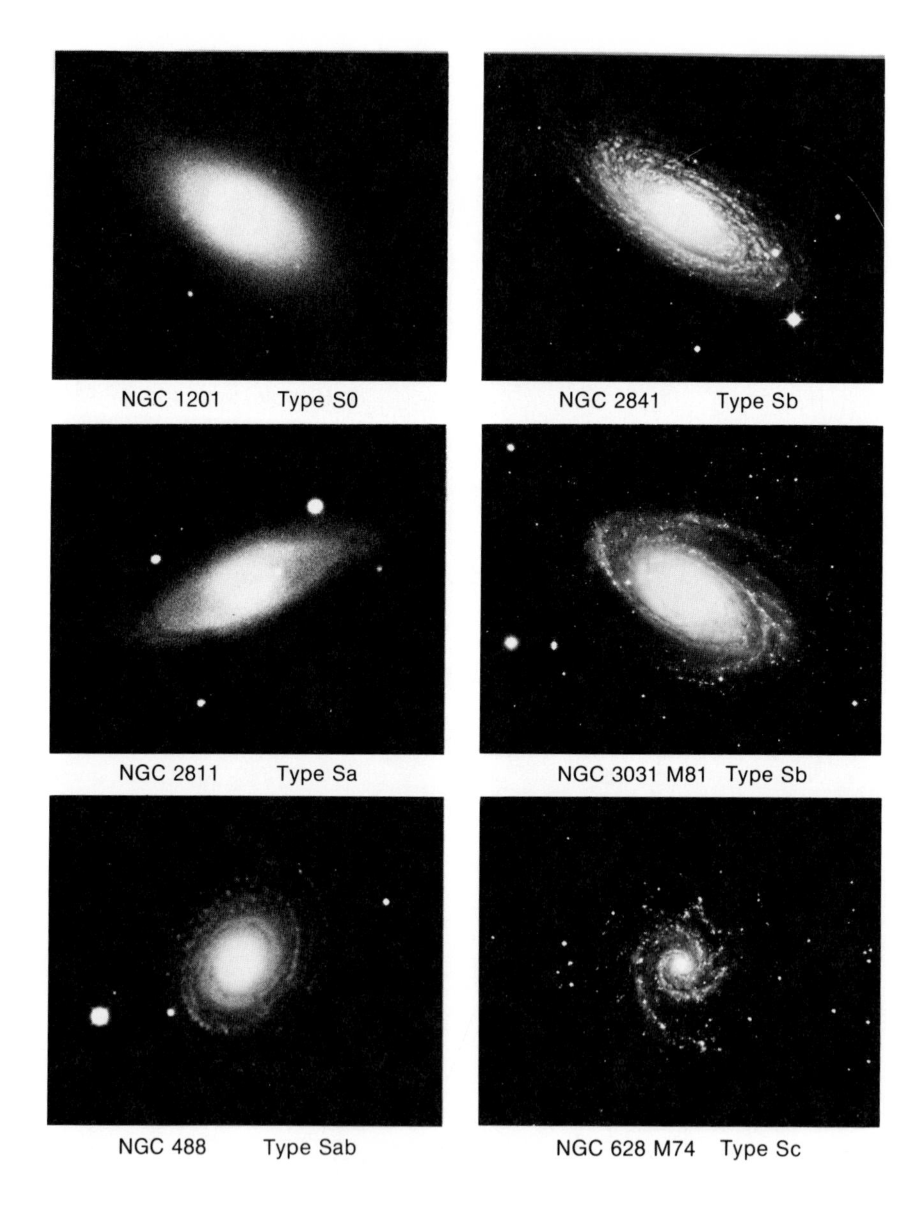

FIGURE 5.7
Examples of the barred types of galaxy.
(Courtesy of the Hale Observatories.)

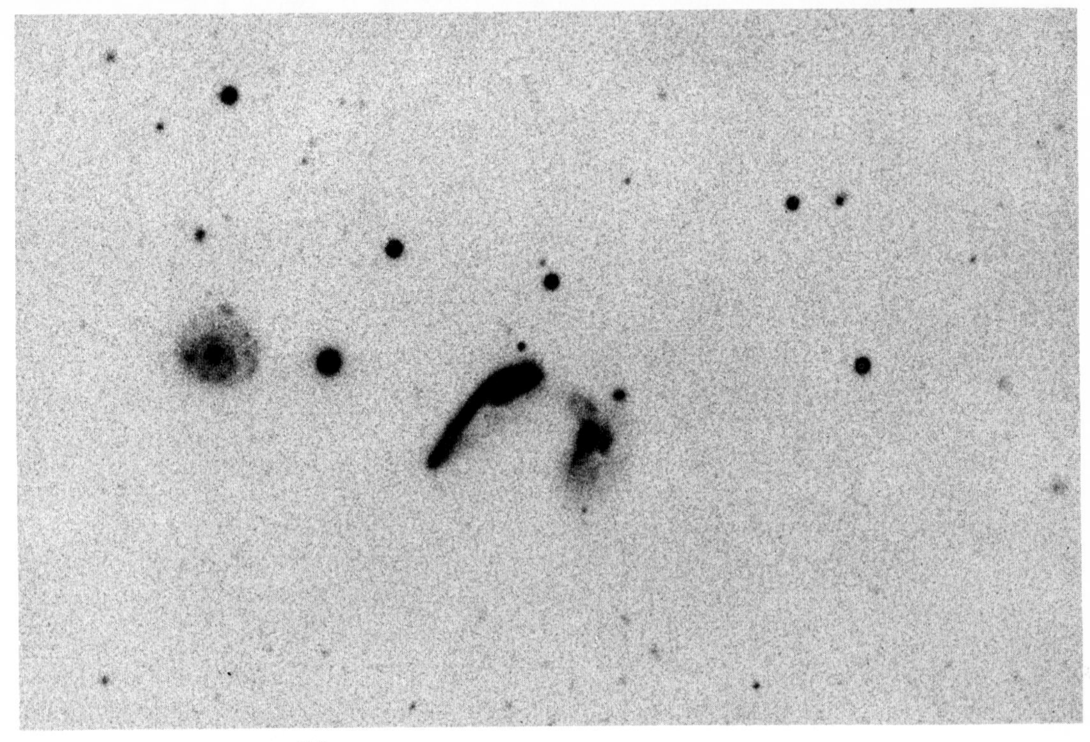

FIGURE 5.8
A photographic negative of a peculiar chain of galaxies. (Courtesy of Dr. H. C. Arp, Hale Observatories.)

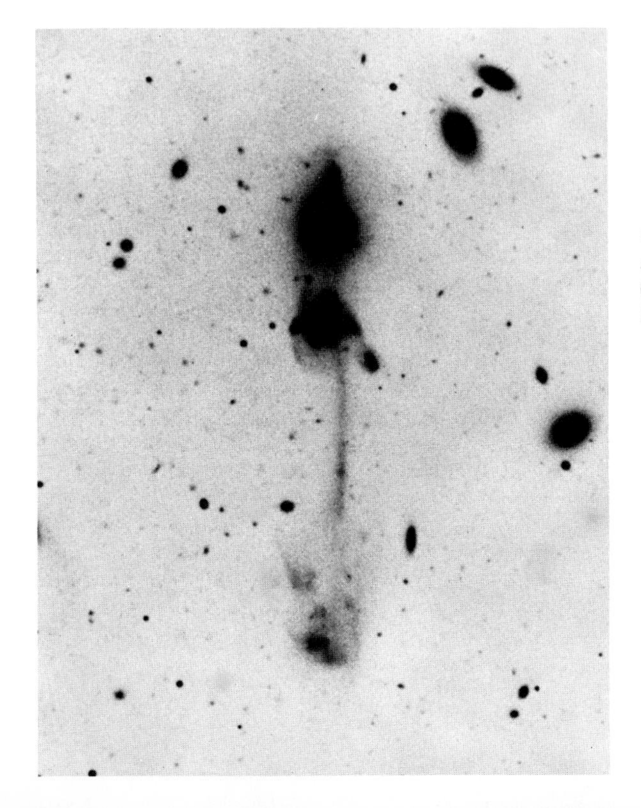

FIGURE 5.9
Another peculiar system, NGC 3561. (Courtesy of Dr. H. C. Arp, Hale Observatories.)

examined closely often show peculiarities. The galaxy M87 appears in Figure 5.4 as a normal globular form belonging to the elliptical class of Figure 5.5. Yet on a short photographic exposure, M87 has the remarkable appearance shown in Figure 5.10. The jet which can be seen emerging from the center is by no means "normal." The light from this jet does not come from stars at all, but from very high-speed electrons moving in a magnetic field. The electrons also emit other forms of radiation, especially radiowaves, and because of this M87 is known as a "radiogalaxy."

The jet of M87 may not be the one-sided affair it looks in Figure 5.10. There could well be an oppositely directed jet which does not happen to emit much visible light. Such oppositely directed jets of high-speed particles are typical of huge outbursts that occur in the central regions of radiogalaxies. When these jets impinge on an external cloud of gas, or on an external magnetic field, they cause an intense emission of radio waves. In Chapter 4 we discussed the explosions of stars, known as supernovae, and we saw that such explosions can cause a star to become temporarily as bright as a whole galaxy. But the outbursts from radiogalaxies are much more energetic than those of supernovae, more on the order of a million supernovae all exploding together. Examples of radiogalaxies are shown in Figures 5.11 and 5.12.

The physical conditions which cause these outbursts of radiogalaxies are exceedingly peculiar. At their centers, many galaxies appear to contain large masses of material, some tens of millions, or even hundreds of millions, times greater than the mass of an individual star. The whole of the material is condensed, probably into a single object, within a region not much larger than our own solar system. Such a region approaches, or perhaps even attains, the extreme conditions of a black hole. From time to time violent motion occurs at, or near, this condensed region, and it is the

FIGURE 5.10
The jet of the galaxy M87. This is a blowup of the extreme central region of the galaxy shown in its entirety in Figure 5.4. (Courtesy of the Hale Observatories.)

FIGURE 5.11
This is a giant elliptical galaxy, out of which a vast cloud of gas and dust seems to have emerged. (Courtesy of the Hale Observatories.)

FIGURE 5.12
One of the most powerful of all radiogalaxies, the system known as Cygnus A. (Courtesy of the Hale Observatories.)

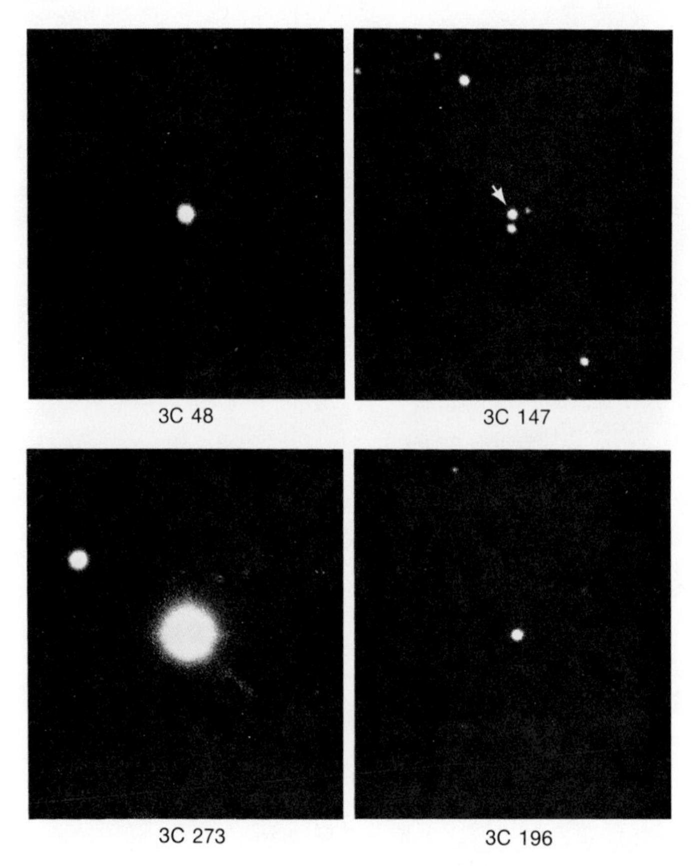

FIGURE 5.13
When photographed directly, the quasars appear like stars. (Courtesy of the Hale Observatories.)

high-speed particles generated by this motion which apparently cause the explosive outbursts of galaxies. A radiogalaxy is simply a galaxy that has experienced such an explosion recently.

Quasars are highly condensed objects of a similar kind. One theory holds that a quasar is actually a galaxy with a particularly active central condensation, a nucleus which has become so inordinately bright that it overwhelms the ordinary starlight of the galaxy. When seen from a great distance, such an object would appear as a central brilliant point of light, as quasars are observed to be. Examples of quasars are shown in Figure 5.13. Notice the quasar

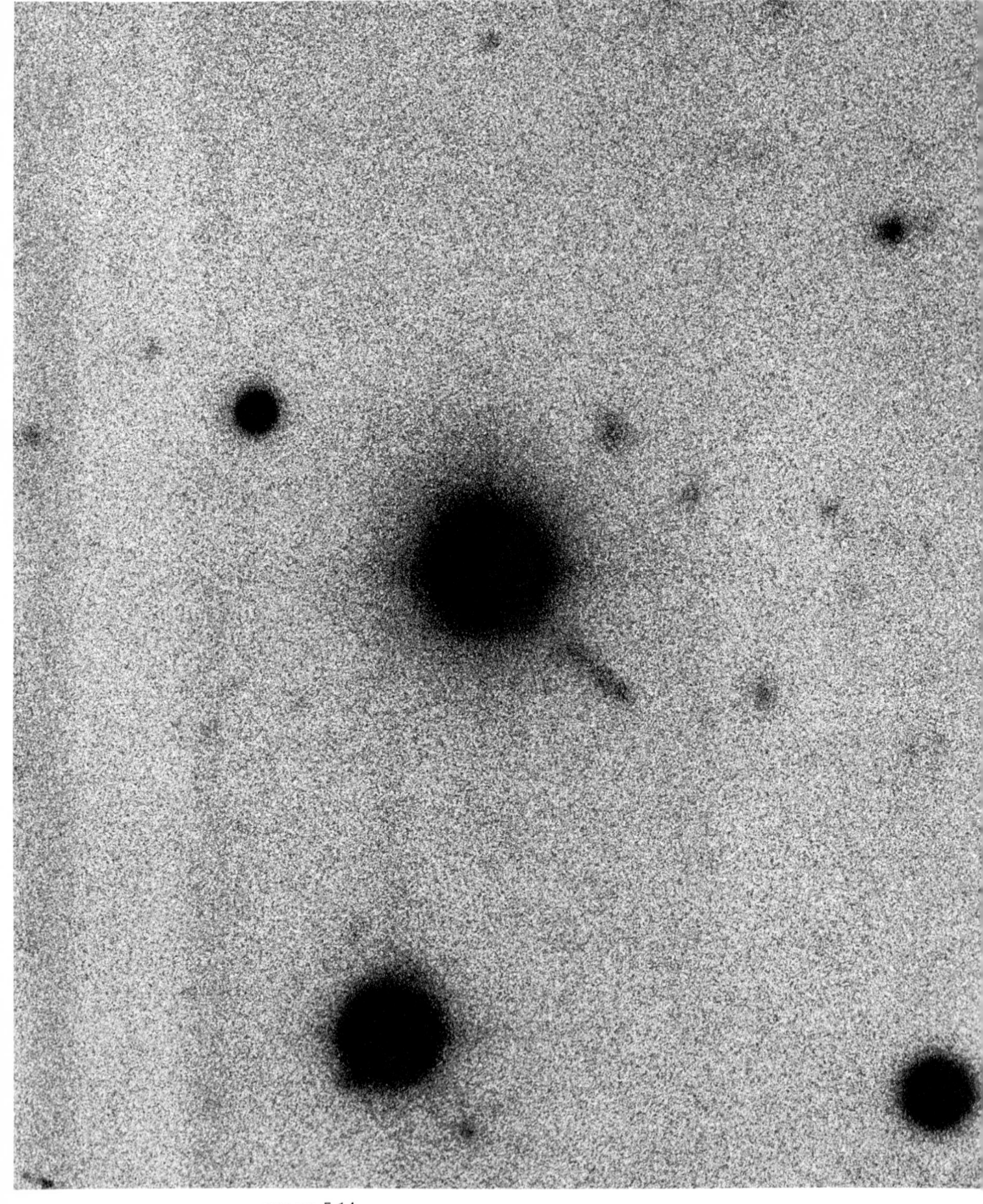

FIGURE 5.14
A long exposure (in negative form) shows the jet of the quasar 3C 273. (Courtesy of the Hale Observatories.)

with the catalogue number 3C 273 has a jet rather similar to that of the galaxy M87. The jet is more clearly seen in Figure 5.14.

Quasars are thought to be the most distant objects visible with a large telescope, indeed, so distant that ten years ago it was believed they might be used as distance indicators to discover the structure of the whole universe. This hope has proved unfounded, however, because of a lack of intrinsic similarity among the several hundred quasars available for study. Only a class of objects that are closely similar to one another can be used as a satisfactory tool for surveying the universe. This essential requirement is better met by certain forms of galaxy than it is by quasars. Unfortunately, galaxies are not luminous enough, hence are not observable at distances great enough, for us to be able to discover the structure of the universe in an unequivocal way. Although for decades astronomers have sought to use the limited region of space and time that they can see in order to discover the properties of the whole universe, results remain uncertain—perhaps inevitably so. At the end of Chapter 2, we saw that an attempt to discover the whole by observing only a part is quite likely a misplaced ambition.

The most important fact for understanding the history of the universe as a whole that has emerged during the last 25 years came about in a quite different way, not from studying galaxies or quasars. In 1965, A. Penzias and R. W. Wilson reported radio waves arriving at the Earth uniformly from all directions in space. This "microwave radiation," with its main intensity in the wavelength range from 1 mm to 1 cm, cannot have been generated by any known type of astronomical object; galaxies and quasars certainly emit radiowaves, but they do not generate an adequate intensity in this particular wavelength range. The microwave radiation is therefore believed to have originated at an epoch of the universe much more remote than any which is observable by the usual astronomical methods. Later in this chapter we shall consider how the astrophysical conditions

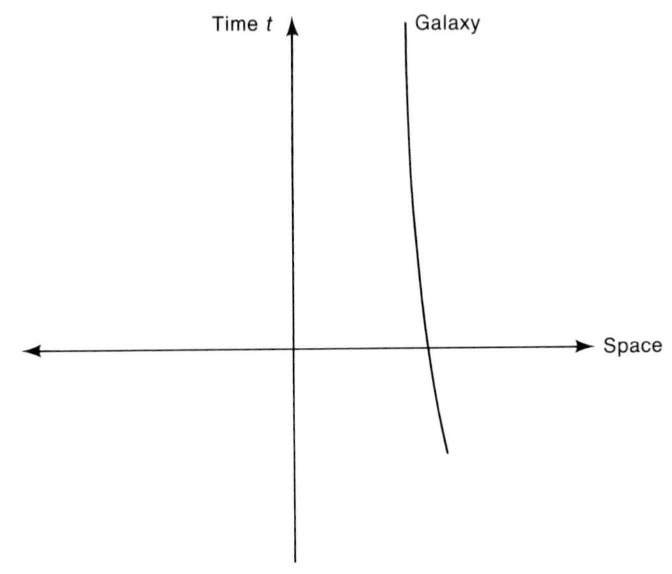

FIGURE 5.15
A galaxy is a cable in four-dimensional spacetime, the cable being approximated here by a line.

inside stars may well be related to such a remote source of the microwave radiation, thereby establishing a connection with the ideas of the preceding chapter.

Out as far as we can readily see—for a distance of about 5,000 million light years—the galaxies appear to be uniformly distributed. Samples of galaxies from different directions are similar to each other; and samples taken at different distances, with allowances made for age variations, also appear similar. Assuming this uniformity extends to the universe as a whole, we can gain an insight into the large-scale motions of the galaxies. Each galaxy is represented in four-dimensional space by a cable whose form represents the motion of the galaxy. Suppose we approximate such a cable by a line, as in Figure 5.15. At any moment of time, the line is characterized by a spatial point. Consider the spatial points, taken at the same moment, for a set of n galaxies, and call them $G_1, G_2, \ldots, G_n$.

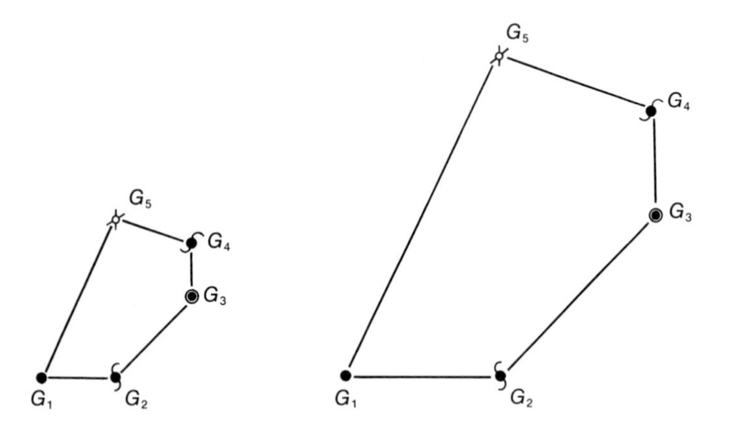

FIGURE 5.16
The uniformity postulate described in the text requires the two polygons, obtained by joining the spatial positions of a set of galaxies at two different moments of time, to have the same shape. The polygons are not required to have the same scale, however. Nor does the uniformity postulate determine whether the larger polygon occurs later or earlier than the smaller one.

A spatial polygon is easily obtained by joining G_1 to G_2, G_2 to G_3, . . . G_{n-1} to G_n, and G_n to G_1. By repeating the procedure for a different moment of time, we obtain two spatial polygons which can be compared. The uniformity postulate for the whole universe requires (and one can prove this by fairly sophisticated mathematics) that the two polygons must have the *same shape*. They can differ, however, in *scale*. The situation for $n = 5$ is illustrated in Figure 5.16. The relationship of the polygons is determined by the relative motions of the galaxies. If the larger polygon occurs later in time, the galaxies are moving away from each other. If the larger polygon is earlier, the galaxies would approach each other as time went on. In the first case, the universe would be expanding, in the second, contracting. The situation revealed by astronomical observation* is that the universe is expanding; so it is the larger polygon of Figure 5.16 that actually occurs later in time.

*Revealed specifically by the redshift of spectrum lines in the light from distant galaxies (for an explanation of this effect, see the next chapter).

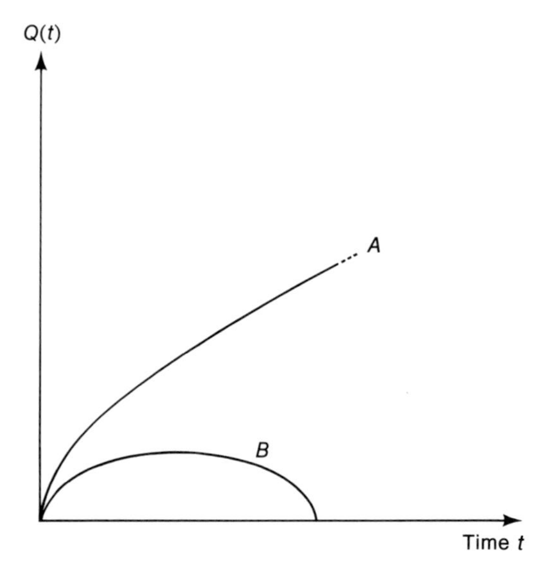

FIGURE 5.17
According to Einstein's gravitational theory, the behavior of the scale factor Q must be either of type A or of type B.

As a matter of notation, it is convenient to characterize the scale of our galaxy polygons by a number Q, with Q changing as the time t changes. With gravitation the only important force operating between galaxies, the curve we obtain when we plot Q vertically and the time t horizontally must be such that the tangent always turns in a clockwise sense. This is because gravitation is a force of attraction—a force of repulsion would cause the curve of Q to turn in an anticlockwise sense.

Einstein's theory of gravitation permits us to go farther than this, but not far enough to discover exactly what the curve for Q must be. We can say that the curve must be one of two types, A or B of Figure 5.17, but we cannot say which. For curve A, the expansion of the universe never ceases. Curve B, on the other hand, has the property that gravitation eventually stops the initial expansion. It does so at the moment when Q attains the maximum of curve B. Thereafter the expansion is replaced by contraction. Observation

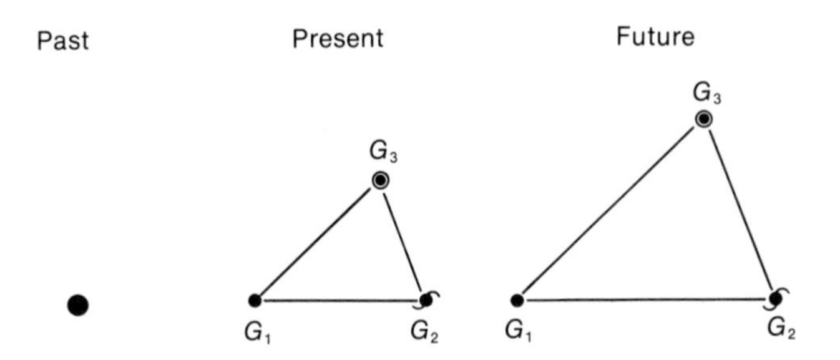

FIGURE 5.18
The triangle formed by joining the spatial positions of three galaxies will be larger in the future than it is now. In the past the triangle was smaller, and far enough back into the past the triangle was collapsed down to nothing at all.

shows that, if the universe is of type B, then we must be living *before* the maximum of Q—i.e., in the expanding phase.

It is common to both A and B that Q was zero at a definite moment in the past. We can see what this result means by returning to our galaxy polygons. In Figure 5.18 we have the simple case of three galaxies, the polygon being then a triangle. Because the universe is expanding, in the future the triangle will be larger than it is now. In the past the triangle was smaller, and far enough back into the past the triangle was collapsed to nothing at all! It is this moment, believed by astronomers to have occurred between 12 and 15 billion years ago, that is usually referred to as the "origin of the universe."

Returning now to the microwave radiation, the origin of the universe is the remote epoch from which the microwave radiation is believed to be derived. The microwave radiation exists because it was there in the beginning—this is the current view of astronomers. In the first few seconds of the history of the universe, the radiation had a temperature of about 10,000 million degrees. The present-day low temperature of only 3 degrees has come about from the expansion of the universe, which has continuously cooled the radiation from its initially very hot state.

It is possible to calculate what abundances the elements would have had as they emerged from this hot early state of the universe. Hydrogen and helium are found to be dominant, with hydrogen atoms about ten times as abundant as helium atoms, in good agreement with the observed relation of helium to hydrogen in cosmic material. The "origin" was thus a kind of springboard from which the universe launched itself. It gave matter the composition that would serve later as an energy source for the stars. Rather like a clock, the universe started by being wound up, and like a clock the universe is running down as the galaxies fly apart, and as the nuclear evolutionary processes within stars approach completion. If the universe is of type *A*, the rundown will be followed through to a state of total inanition. If it is of type *B*, the universe will eventually fall back on itself, going out of existence (when Q returns to zero) just as mysteriously as it came into being at the moment of its origin.

FURTHER READINGS

F. Hoyle, *Astronomy and Cosmology*. W. H. Freeman and Company, 1975.

E. P. Hubble, *The Realm of the Nebulae*. Yale University Press, 1936.

A case for regarding the universe as being of type *A* (open), not type *B* (closed), has recently been given by J. R. Gott III, J. E. Gunn, D. N. Schramm, and B. M. Tinsley, "Will the Universe Expand Forever?" *Scientific American*, **234** (March 1976), 62.

6

THE ORIGIN OF THE UNIVERSE

GEORG FRIEDRICH BERNHARD RIEMANN
(1826–1866)

(Photo courtesy of
Foto Schubert, Göttingen.)

6

THE ORIGIN OF THE UNIVERSE

Our awareness of the world includes both local experience and the far-ranging observations of the astronomer. In Chapter 5 we arrived at a most remarkable conclusion. We saw that the world taken in this sense has had only a finite past history, that it all began about 15 billion years ago. The question we have to ask in the present chapter is whether the world we are aware of constitutes the whole universe. Let me state my view at the outset. I do not believe that the whole universe suddenly came into existence out of nothing. Rather, in this chapter I seek to develop the idea that the world of our experience sprang into existence 15 billion years ago in a sudden transition from another existence, out of another world.

Science progresses by extending the territory over which its theories hold good, as we saw at the end of Chapter 2. The theory discussed in Chapter 5 applied only to the world we are now aware of. Call this theory *T*. Our aim here will be to find a wider theory, *T′*, that applies equally well to our world, but that can be extended to include an existence which preceded our own.

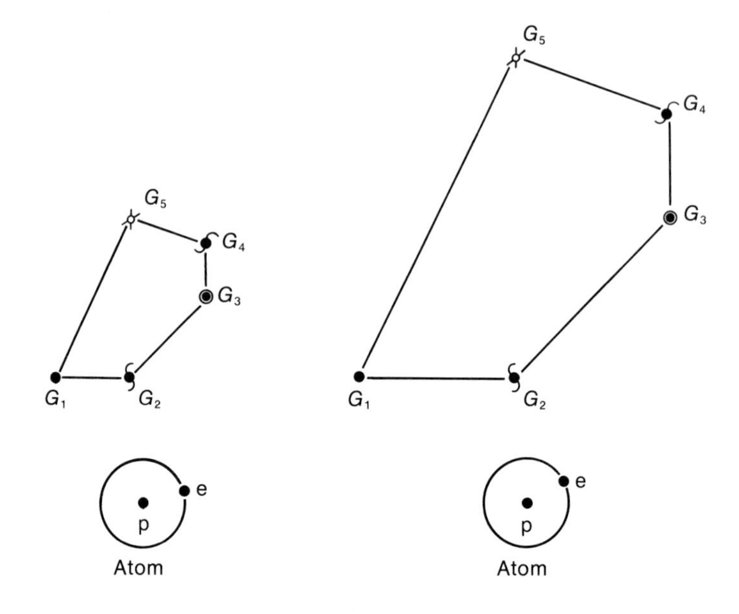

FIGURE 6.1
The scale of the galaxy polygon increases if the unit of measurement is derived from atoms that are fixed in size.

The unit of distance used in the theory T is set by the size of a particular atom, hydrogen, for example. This leads to the familiar situation of Figure 6.1, with the distances between galaxies increasing as time goes on. Instead of continuing with this procedure, suppose we choose the average separation of galaxies as our distance unit, keeping the galaxies at all time at a unit distance apart. For consistency, the sizes of atoms must then decrease as time goes on, as is indicated in Figure 6.2. By giving mathematical expression to this second way of relating galaxies and atoms, we arrive at a new theory, T'. The method is the following.

The size of an atom is determined by the masses of its constituent particles—the larger the masses, the smaller the atoms, the relation being reciprocal. To understand how masses might change, we regard the mass of a particle as a property which it derives from interactions with other particles, in the manner of Figure 6.3. Then

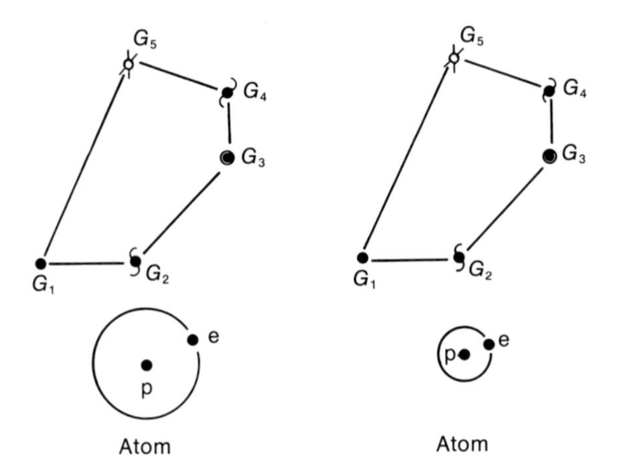

FIGURE 6.2
As an alternative to Figure 6.1, keep the galaxy polygon fixed; then atoms decrease in size with time.

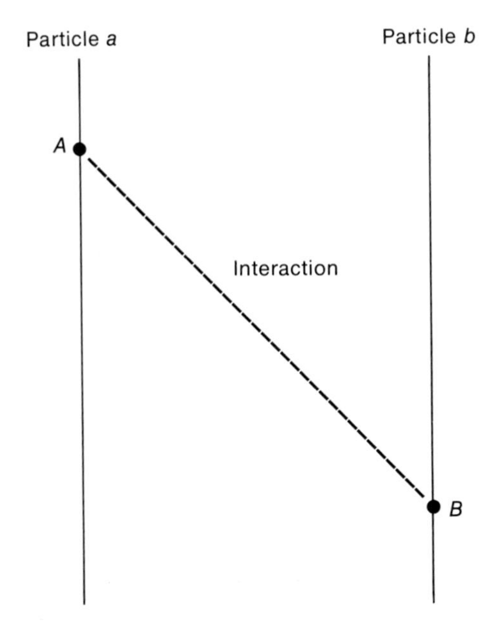

FIGURE 6.3
The mass of particle *a* at the point *A* is to be regarded as made up of contributions from other particles. Here we have the contribution from a typical point *B* on the path of particle *b*.

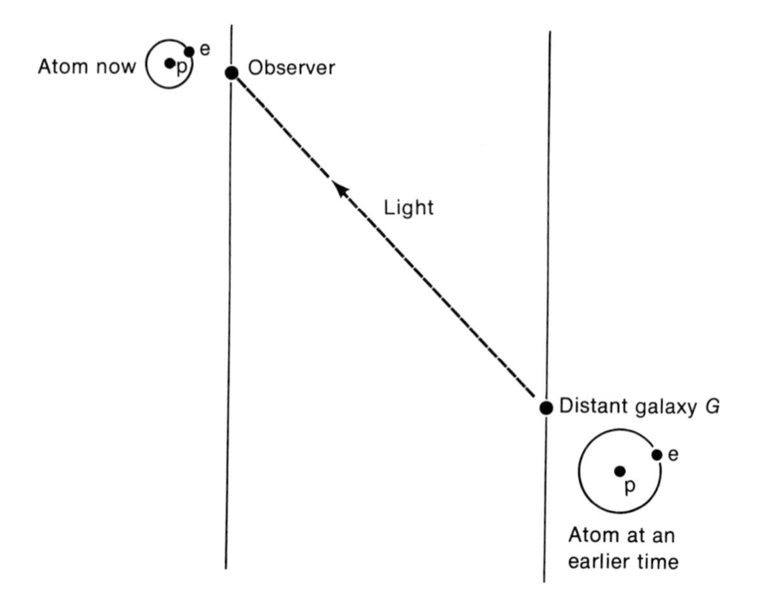

FIGURE 6.4
Light from a distant galaxy was generated at a time when atomic masses were less than at present. For this reason, radiation from an explicit atomic transition is redshifted with respect to present-day radiation from the same transition.

to establish T', we must express the interaction of Figure 6.3 mathematically in such a way that Figures 6.1 and 6.2 become precisely equivalent to one another for all aspects of the world that we are aware of.*

As an example of this equivalence, it is interesting to consider how the new T' explains the redshift of the light from a distant galaxy, *G* in Figure 6.4. Because the light is emitted at an earlier time, the distant hydrogen atoms (for example) have a smaller mass than local hydrogen atoms. The light emitted from *G*, which we are now

*For the mathematical details of how this is actually done, see F. Hoyle and J. V. Narlikar, *Action at a Distance in Physics and Cosmology* (San Francisco; W. H. Freeman and Company, 1974).

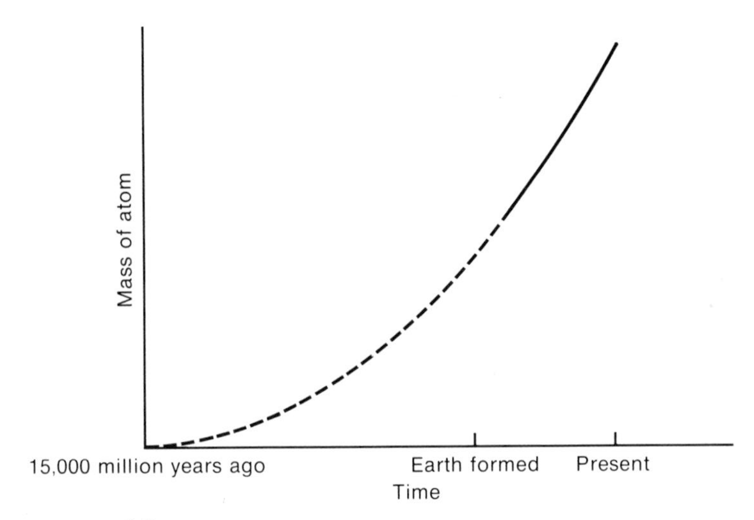

FIGURE 6.5
The behavior of the mass of the hydrogen atom with respect to time. The solid part of the curve is observed, the dotted part inferred.

receiving in our telescopes, will therefore be different from light emitted by hydrogen atoms in a terrestrial laboratory. In fact, the light will be redder, because the frequency values of the light emitted by the distant atoms and by the local atoms are proportionate to their respective masses. Actual observations of this redshift effect, so far as they extend, are in good agreement with the calculated mass curve of Figure 6.5. The curve goes back to a moment of time when the mass was zero, nothing at all. This is the "origin" of the universe once again, showing itself in a different guise in the theory T'.

Figure 6.6 is an illustration of the world we are aware of according to T'. The separations of the galaxies remain fixed here, since in T' the galaxies at all times remain a unit distance apart. When our world is represented in this way, the extension to Figure 6.7 appears as an almost self-evident step. The time $t = 0$ in Figure 6.7 is the moment of transition between our world and a kind of mirror-image

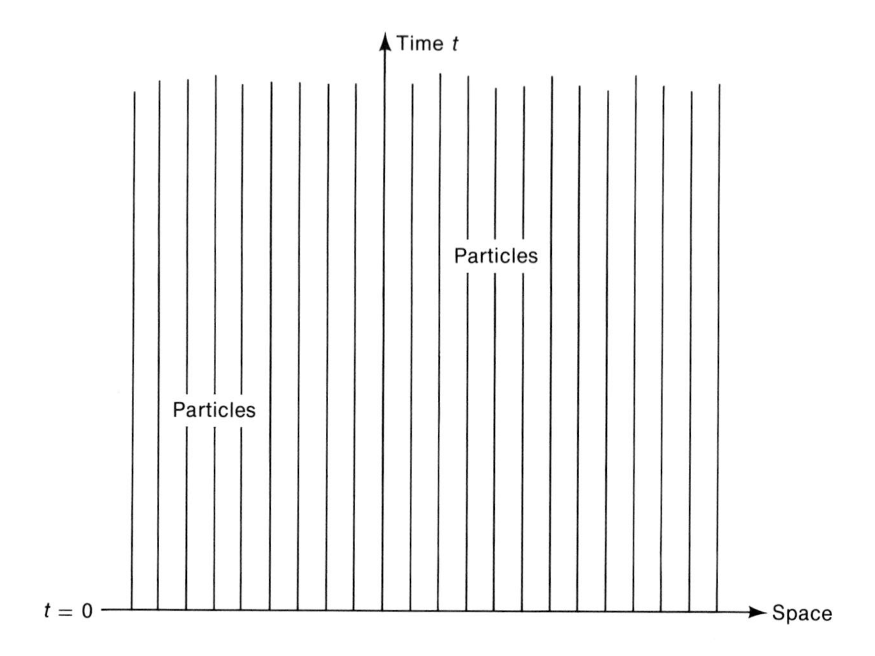

FIGURE 6.6
The lines representing the galaxies are all parallel to the time axis.

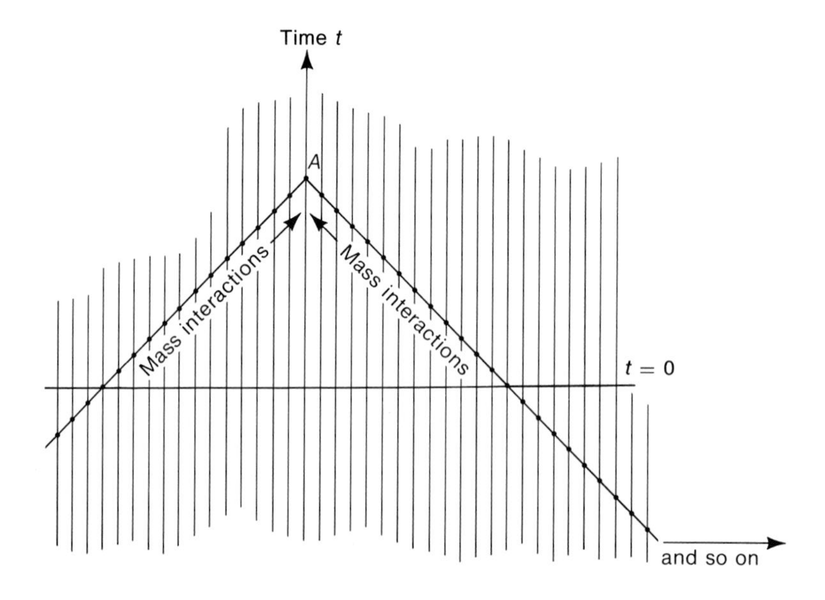

FIGURE 6.7
Simply extending the paths of the particles backward in time leads to difficulties.

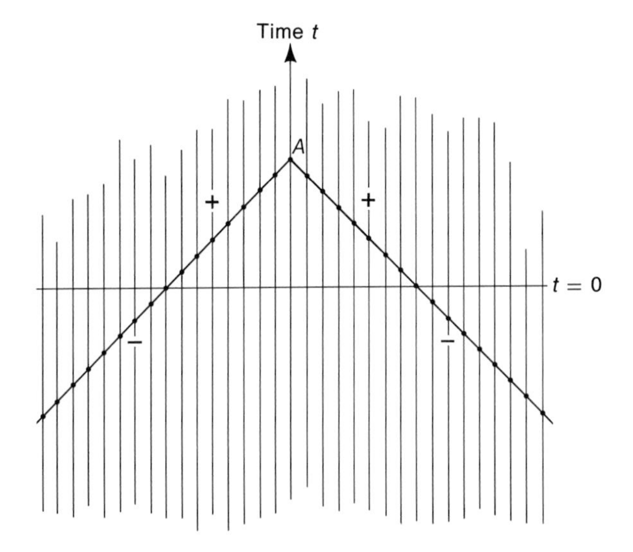

FIGURE 6.8
The difficulties are overcome, provided the moment $t = 0$ is taken to separate positive and negative contributions to the masses of the particles.

world existing before $t = 0$. Yet this extension is by no means as trivial as it might seem, since the mass curve of Figure 6.5 must still follow from the interaction picture of Figure 6.3, otherwise T' would not be the same as T within the range of our own observations. This further problem turns out to be elegantly solved by the scheme of Figure 6.8, in which interactions preceding $t = 0$ contribute negatively to the mass.* Hence we have reached our goal: we discover an existence preceding our own existence.

---

*There are two ends to the interaction of Figure 6.3, an end at each particle. For an end falling into our world (time later than $t = 0$) the interaction is multiplied by $+1$; for an end in the preceding world (time earlier than $t = 0$) the interaction is multiplied by -1. This rule has a symmetric effect between the two halves of Figure 6.8. Whenever the two ends of an interaction fall on the same side of $t = 0$, the product of the two interaction factors is $+1$, since $(+1) \times (+1) = (-1) \times (-1) = +1$. When the two ends of an interaction fall on opposite sides of $t = 0$, the product of the two interaction factors is -1, since $(+1) \times (-1) = -1$.

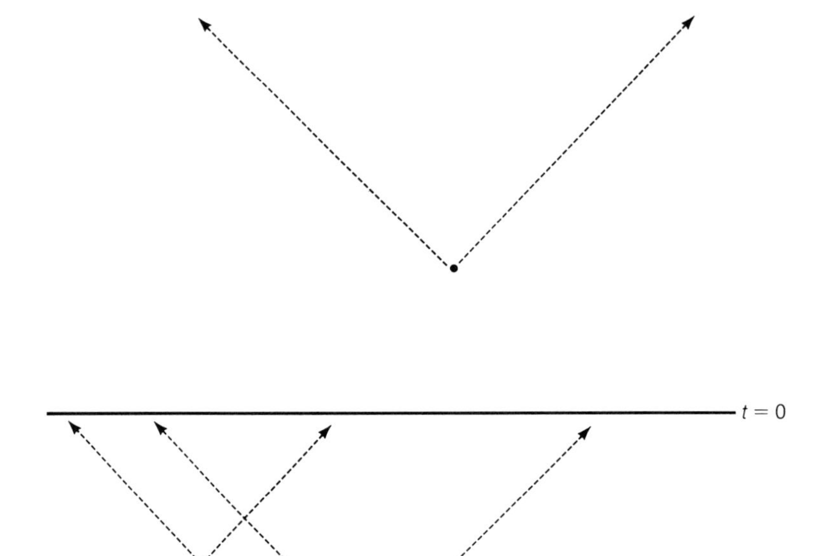

FIGURE 6.9
Radiation propagates in the directions of the dotted lines, leading to the origin of what is known as the "microwave background."

Suppose light travels always in the same time sense, as in Figure 6.9, and suppose that galaxies and stars exist on the other side of $t = 0$ just as they do on our side. Starlight generated on the other side propagates toward $t = 0$. What happens to this radiation? Can it come right through $t = 0$, and so permit us to observe the galaxies on the other side? The answer to this question is, no. The starlight is powerfully absorbed near $t = 0$, because near $t = 0$ the particles of gas lying between the stars have very small masses, and particles of small mass are exceedingly powerful absorbers of radiation, very much more so than the particles of our everyday environment. Such particles are also powerful emitters of radiation, and they immediately reemit the absorbed radiation. In a sense, therefore, radiation does indeed come through $t = 0$. However, the galaxies and stars on the other side are entirely blurred out by the absorption and re-emission process, and so cannot be observed at all in the usual sense. Nevertheless, the blurred radiation crossing through $t = 0$ from the other side should be observable, and it is in fact detected as the microwave radiation discovered by Penzias and Wilson.

We are well-used to the idea that information can be carried by radiation; indeed, the transmission of information forms the whole basis of radio broadcasting. The flow of information always goes in the time-sense of propagation of the radiation, from past to future, never in the opposite time-sense. The sequence of cause and effect goes the same way. Figure 6.9 shows that the past-to-future flow of information, and of cause and effect, comes from the other side of $t = 0$ into our side. It is relevant, therefore, to ask if certain of the events observed by astronomers could have been caused by this flow. Investigation shows that stars existing before $t = 0$ would persist throught $t = 0$, emerging as starlike condensations on our side. The structural forms of the galaxies could indeed be the result of information flowing to our side across $t = 0$. Whole showers of stars can come through $t = 0$, giving a considerable measure of control from the other side to our side. Many conditions in our world could well have been created by the different world which preceded us.

In conclusion, let us consider more closely the particular moment $t = 0$. All mass interactions switch their sign at this moment. How did this well-ordered transition come about? To answer this

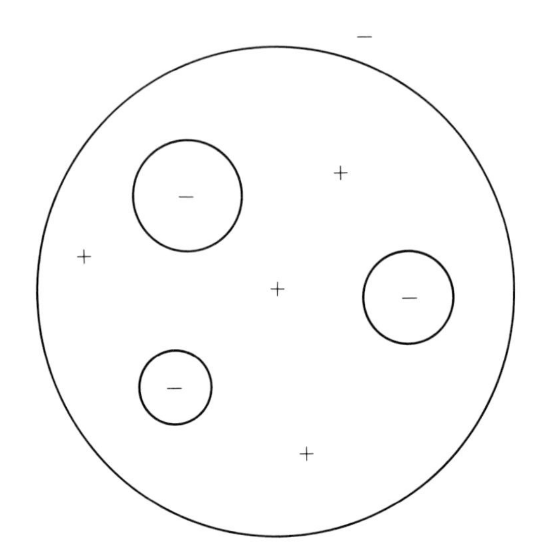

FIGURE 6.10
A schematic representation of large-scale plus and minus aggregates.

question, consider the situation of Figure 6.10, in which there are plus and minus aggregates of quite general shapes. Figure 6.10 covers a territory much larger than the portion of the universe accessible to practical observation. Indeed, our observations of all the galaxies, even the most remote ones visible in the largest telescopes, would occupy only a comparatively small element of just one of the aggregates of Figure 6.10—for definiteness, let us say a plus aggregate.

The property distinguishing the two kinds of aggregate is simply the sign of the mass interaction. Particles in a plus aggregate contribute positively, and particles in a minus aggregate contribute negatively, to the mass interaction. The effect on a particle anywhere in the universe will, in this picture, be a complicated addition of contributions from the particles in all the aggregates. If we make the sensible assumption that minus aggregates are as important as plus aggregates, the addition of all interactions at an arbitrary place is as likely to be negative as positive. Regions where the contributions add to a positive total will be separated from regions with a negative total by surfaces on which the plus and minus contributions just cancel each other. The mass of a particle at a point on such a surface will be zero.

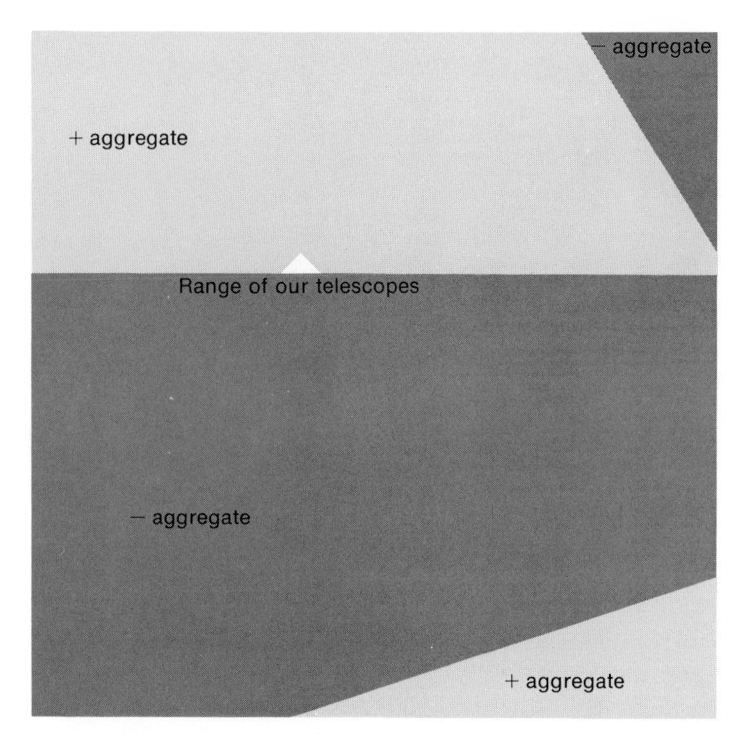

FIGURE 6.11
The range of our telescopes, represented schematically here by the white triangle, may reveal only a tiny element of the entire universe.

In a quite general way, we now arrive at an understanding of the moment $t = 0$. This moment defines a portion, perhaps only a small portion, of one of the surfaces of Figure 6.10. The essential lesson we learn from the study of the motions of galaxies is that our element of the universe, the world we are aware of, happens to lie near a surface of zero mass. The surface in question need not be a single unique surface, as we formerly took it to be. To emphasize this point, let us redraw Figure 6.10 to show that the range of our astronomical observations may cover only a small element of a much vaster universe. This we can see in Figure 6.11, where we have the concept of a universe much greater in scale than any which has hitherto been considered in astronomy.

7

NOBODY'S UNIVERSE

JOHANN CARL FRIEDRICH GAUSS
(1777–1855)

(Photo from Historical Pictures Service, Inc.)

7

NOBODY'S UNIVERSE

The key to the extension of the universe, illustrated in Figure 6.11, lay in the mass interaction of Figure 6.3, and in the zero-mass surfaces of Figure 6.10. The interactions were considered to operate over great distances, on the scale of the aggregates of Figure 6.10. We have gained this insight, however, at the cost of coming into conflict with the views of most physicists, who hold that the universe in the large simply dances to the tune called by mathematical rules of local origin, a view that must break down if the fundamental property of "mass" were derived from interactions ranging over the whole universe.

Rather than abandon their point of view, I suspect most physicists would prefer to accept the alternative position, that the whole universe popped into existence out of nothing, which I rejected at the outset of Chapter 6. The physicist fears that, once the door is open to nonlocal influences on the physical laws, there is no telling how far things might go. They could possibly affect all the physical

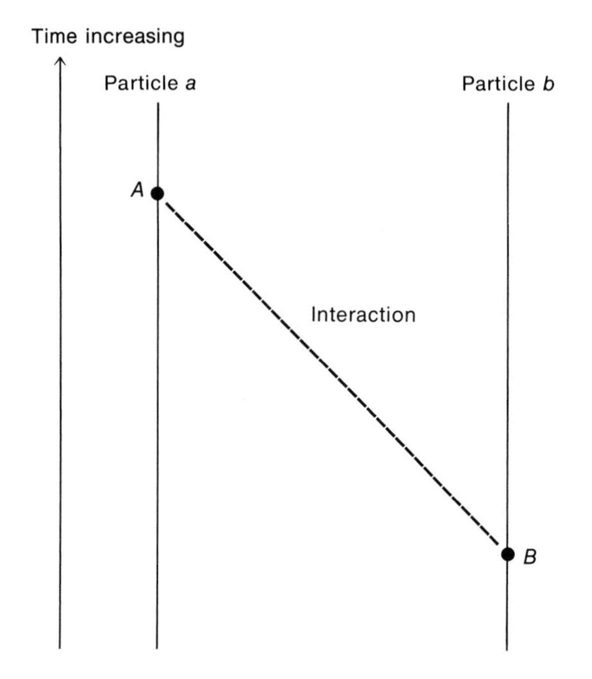

FIGURE 7.1
The point *A* is at a later time than point *B*. In Maxwell's theory the electrical interaction goes from *B* to *A*, not from *A* to *B*.

interactions. I propose here to discuss this unpopular but subtle issue still further, beginning with a return to Maxwell's theory of the electrical interaction.

In Maxwell's theory there is no reason why in Figure 7.1 point *A* on the path of particle a should not influence point *B* on the path of particle *b*. Yet in Figure 7.1 point *A* is at a later time than point *B*; so an interaction from *A* to *B* would invert the usual time-sense, permitting events to occur before their causes, for instance, as if an electric light in the home should begin to shine before we press the appropriate switch. Of course, Maxwell's theory also permits the normal sequence of cause and effect, past-to-future, with an interaction going from point *B* to point *A* in Figure 7.1. The situation is that Maxwell's theory of the electrical interaction permits *either* time sequence to occur, either past-to-future or future-to-past.

Herein lies a crucial objection to the local view of physics—for unless a theory entirely explains the observed phenomena, it must surely be judged incomplete. Maxwell's theory is incomplete because it does not explain the particular time-sequence of cause and effect that we actually observe in the world. In all the developments of local physics that have occurred since Maxwell, this situation has not changed.

The footnote on page 45 of Chapter 3 pointed out that the great German mathematician Carl Friedrich Gauss failed in his attempt to formulate the electrical interaction for a far deeper reason than his contemporaries. Gauss's attempt led to the situation that, if point *B* in Figure 7.1 influences point *A*, then point *A* must of necessity influence point *B*. The electrical interaction was required to go both ways, both from past-to-future and from future-to-past. Subsequent attempts to follow Gauss's line of reasoning have all led to this conclusion.

The situation continued unchanged until the 1940's, when John Wheeler and Richard Feynman put forward a remarkable new idea. They boldly accepted Gauss' idea of a *local* electrical interaction going equally future-to-past and past-to-future. They then argued that the observed asymmetry of cause and effect in the world, past-to-future only, was a *nonlocal* effect arising from the large-scale influence of the whole universe. We are well-used to the situation of Figure 7.2, in which a local system is subject to influences from the past—the astronomer makes use of this kind of influence whenever he observes distant stars and galaxies. What we are not used to is the situation of Figure 7.3 (page 127), in which a local system is subject to influences coming from the future. Yet, if the basic physical laws go equally future-to-past and past-to-future, Figure 7.3 must be just as applicable as Figure 7.2. Wheeler and Feynman referred to the influences of Figure 7.3 as the "response" of the universe. They found that the response of the universe could

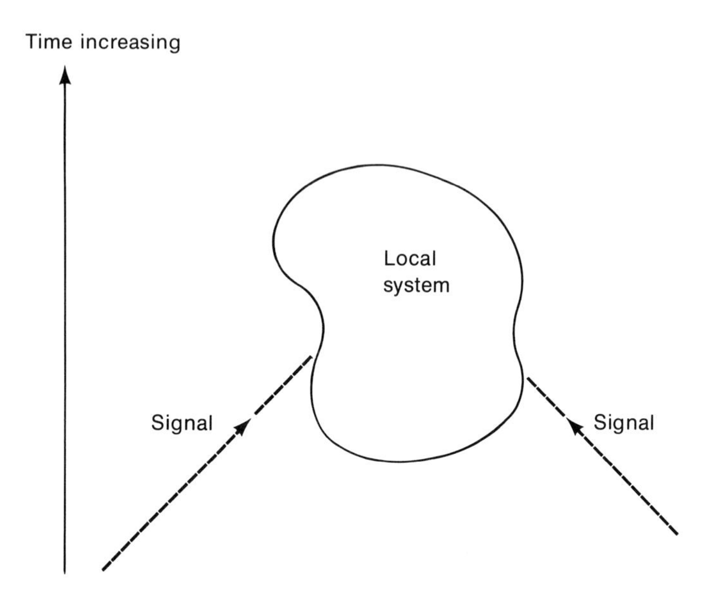

FIGURE 7.2
The local system is influenced by interactions coming from the past.

exactly cancel local future-to-past effects. Local past-to-future effects survived, however, and were in fact doubled. Thus the idea of Wheeler and Feynman led to the following relationship:

$$\text{Time-symmetric local laws} + \text{Response of the universe}$$
$$\Rightarrow \text{Observed sequence of cause and effect.}$$

According to this relationship, the influence of the universe is essential for understanding the normally experienced relation of cause and effect. Purely local laws in themselves are not sufficient.

It is, of course, one thing to conceive this idea in a general way and quite another to make it work in precise mathematical terms. What Wheeler and Feynman showed was that the idea works successfully for prequantum physics. The question remained of whether the idea could also be made to work within the framework of quantum mechanics. This particular problem tormented me personally for close on to ten years. in the end, my colleague Jayant Narlikar and I came to the conclusion that indeed the idea works just as well in quantum mechanics as it does prequantum mechanics.*

This development turned out to have relevance for the philosophical issues concerning quantum mechanics that were discussed in Chapter 3. We saw that the tapestry of cables and threads possesses a greater freedom of form than the scientists of the nineteenth and early twentieth century believed, especially in the detailed behavior of the threads which represent the basic particles from which matter is constituted. Such particles have many possible paths, for which probabilities can be calculated by the method of quantum mechanics. This circumstance led us to ask: Is there an actual universe corresponding to each form of tapestry? Or is there just one universe, and, if so, what is meant by the other possible forms of the theory?

To come to grips with these questions, consider what is said in the usual formulation of quantum mechanics. Repeat an experiment involving the threads of the tapestry, i.e., involving electrons, protons, neutrons, . . . , and do so under similar circumstances. The same result will not be obtained in each case; the results will vary from case to case, covering the various possibilities permitted by quantum mechanics. That is to say, the distribution of the results will reproduce the probabilities calculated from the rules of quantum mechanics. So quantum mechanics gives a correct statistical distribution for such a repeated experiment, although it does not say anything about the outcome of just one of the experiments.

*The basic mathematical condition is given in equation 124, page 80, of *Action at a Distance in Physics and Cosmology* (San Francisco: W. H. Freeman and Company, 1974).

In practice, we are mostly concerned with systems that contain many particles; hence it is the statistical behavior of the particles in which we are usually interested. For example, if we wish to study the emission of light and heat from the Sun, we are not so much interested in what one particular atom happens to be doing, but in what a very large number of atoms are doing. For a very large number, statistical information is fully adequate. Operationally, this is the way quantum mechanics is used to calculate the behavior of physical systems.

The statistical interpretation of quantum mechanics implies that each individual experiment gives a definite result, of which we are aware as soon as the individual experiment is actually performed. Suppose a cat is fastened in a box with a gun pointed at its head. A device is arranged to pull the trigger according to whether a certain kind of electrical signal is received by the trigger mechanism. Let the signal depend on whether or not a neutron within an unstable atomic nucleus changes into a proton (when this happens the nucleus emits an energetic electron, which can be used to generate an electrical signal). A definite number of these unstable nuclei, say, N, is used in the experiment. The trigger mechanism is activated only for a suitably short time Δt. The chance that one of the nuclei undergoes change in this time Δt is obtained by dividing the product $N\Delta t$ by the 'lifetime' λ of the nucleus, $N\Delta t/\lambda$. Choose Δt so that this chance is, say, $\frac{1}{10}$. If the experiment is repeated many times, each time with a different cat, $\frac{1}{10}$ of all the cats will be shot. But in any one experiment there is no means of knowing ahead of time whether or not the particular cat involved in that experiment will be shot. Thus from quantum mechanics we can say ahead of time that $\frac{1}{10}$ of all cats will die, but we have no means of predicting which individual cats will be the unfortunate ones. Yet if such experiments were actually done, the decisions would be definite—a particular cat would either die or it would not die. Given a doomsday radioactive bomb, a similar device could be used to destroy, or not to destroy, all life on the Earth.

Usually the paths of the large cables in our tapestry depend only statistically on their many constituent threads. The paths of the large cables are then calculable definitively from quantum mechanics, and are no different from the paths calculated in prequantum physics. Yet the examples of the previous paragraph show that this need not always be the case. In the cat experiment, the fate of a creature of substantial complexity turned on a single unpredictable event. Extension of the cat experiment to the doomsday bomb showed that even the fate of all life on the Earth could turn on a single unpredictable event—the paths of the large cables in the tapestry can, in exceptional situations, turn on a single one of the vast number of threads woven into the tapestry. When a human makes a "crucial" decision, something like this probably happens within the cable that constitutes the human brain. The ability to "change one's mind" may well be the same thing again. So may be a "stroke of genius." These human experiences have an inherently unpredictable quality which may reflect the unpredictable nature of a single quantum event.

Many persons have always felt that physics should be just as definite as the universe itself, that it should be possible to go beyond the prediction of death for $\frac{1}{10}$ of the ensemble of cats to an actual statement of *which* cats would be the unlucky ones. This strong subjective feeling, that the definite outcome of an individual experiment should be predictable, was countered some forty years ago by the mathematician John von Neumann, who sought to demonstrate the *impossibility* of making such individual predictions. Von Neumann's proof* is still considered today to be the

---

*Originally published in *Mathematische Grundlagen der Quantenmechanik* (Berlin: Springer, 1932).

crucial objection to such an extension of the ideas of quantum mechanics. Yet the fact that by observation we can actually *do* what von Neumann claimed to be impossible is most disturbing. By observation we can discover the identity of the cats that die, although by calculation such discovery is apparently impossible. This, indeed, is a mysterious aspect of quantum mechanics. How can the human brain, seemingly only a sophisticated form of computer, do what has been proved mathematically to be impossible?

Let us see if we can use the concept of the response of the universe to cast light on this seemingly self-contradictory situation. The equivalence set out above can be rewritten in the form:

Local laws + Response of universe
$\Rightarrow$ Predictions of the outcome of experiments.

Taking the local laws to be formulated with time-symmetry—i.e., without any *ad hoc* assumption of the time-sequence of cause and effect—it turns out that not even statistical predictions can be made unless the response of the universe is included in the equivalence. From the local laws alone it is not possible, for example, to define the concept of the lifetime of an unstable nucleus. Hence the cat experiment is not meaningful within the context of only local laws. We are therefore required to consider the situation already indicated diagrammatically in Figure 7.3. All the usual results of quantum mechanics are then obtained, provided the response of the universe takes an appropriate form, the form referred to in the footnote on page 123. But suppose now that a statistically adequate form of the response of the universe is not the whole story. Suppose the response of the universe is not the same in detail for each of the cats in our experiment. Suppose the response of the universe contains variations from one cat to another, and suppose it is these

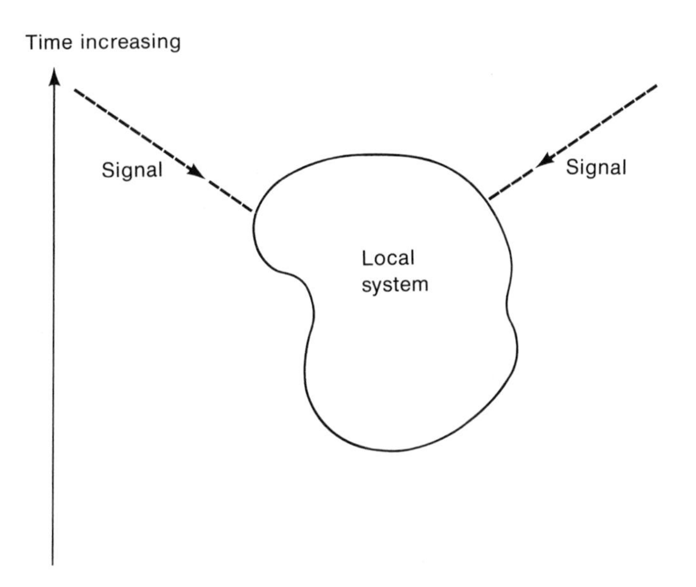

FIGURE 7.3
A local system influenced by interactions from the universe coming from the future.

variations that decide which cat dies and which does not. In short, when we think we are setting up similar experiments, we are not really doing so. The local situation may be entirely the same for the different cats, but the response of the universe is not the same. Is there anything in von Neumann's proof to rule out such an idea? Examination of the proof shows it to be designed for local systems only. The proof fails when variables outside the local system become important, as they are in the situation of Figure 7.3.

Our projected enterprise of calculating the outcome of a single experiment still fails in practice, because at present we know only the statistically correct form for the response of the universe. But this is not to say that we are forever debarred in principle from making such a calculation. Success in the enterprise would require details of the distant universe to be included in the calculation. So far we do not know how to make such an ambitious calculation.

Success may come one day, however, but only from a nonlocal form of physics, the kind of physics that is not at all popular right now.

Although we cannot as yet give a mathematical form to the detailed variations in the response of the universe, it seems that we are operationally aware of such details in our conscious experience. When we look at the ensemble of cats, we are aware of the identities of the unlucky ones. By looking, but not by calculation, we are aware of the effects produced by the precise form that the response of the universe takes in each individual case. It is an interesting speculation that consciousness itself may arise from this interaction of our mental processes with the universe in the large—not the usual cause-and-effect interaction of Figure 7.2, but the interaction with the future shown schematically in Figure 7.3. This speculation appears, almost inevitably, when we seek to relate local physics to the structure of the whole universe. Our experience then contains a major environmental component by means of which the phenomenon of consciousness may well arise.

FURTHER READINGS

F. Hoyle and J. V. Narlikar, *Action at a Distance in Physics and Cosmology.* W. H. Freeman and Company, 1974.

C. F. Gauss, *Werke,* V (1867), 629.

J. A. Wheeler and R. P. Feynman, "Interaction with the Absorber as the Mechanism of Radiation," *Reviews of Modern Physics,* **17** (1945), 157.

J. A. Wheeler and R. P. Feynman, "Classical Electrodynamics in Terms of Direct Interparticle Action," *Reviews of Modern Physics,* **21** (1949), 425.

8

THE GEOPHYSICIST'S UNIVERSE

ALFRED LOTHAR WEGENER
(1880–1930)

(Photo from Historical
Pictures Service, Inc.)

8

THE GEOPHYSICIST'S UNIVERSE

Two views of the Earth, taken from space during the Apollo Missions of the National Aeronautics and Space Administration (NASA), are shown in Figures 8.1 and 8.2. It is interesting to take a close look at Figure 8.1. The vast land mass of Africa can be seen at the upper left, connected to Arabia at top center. At the upper right, a swirl of cloud is a storm in the Indian Ocean. Clouds show white, and they tend to be curved in spirals, an effect caused by the spin of the Earth.

The white mass at the bottom of Figure 8.1 is not cloud, however, but ice. This is the Antarctic, the scene of Captain Scott's heroic walk to the South Pole. When one looks carefully, it is not hard to see where the ice ends. At its boundary, the Antarctic ice meets the dark, forbidding southern oceans, which navigators of the eighteenth century tried to penetrate. It was in these waters that the whaling industry of Herman Melville's *Moby Dick* became established in the nineteenth century.

FIGURE 8.1
Apollo 17: A view of the Earth on the journey toward the Moon. (Courtesy of NASA.)

FIGURE 8.2
Apollo 13: Another view of the Earth. (Courtesy of NASA.)

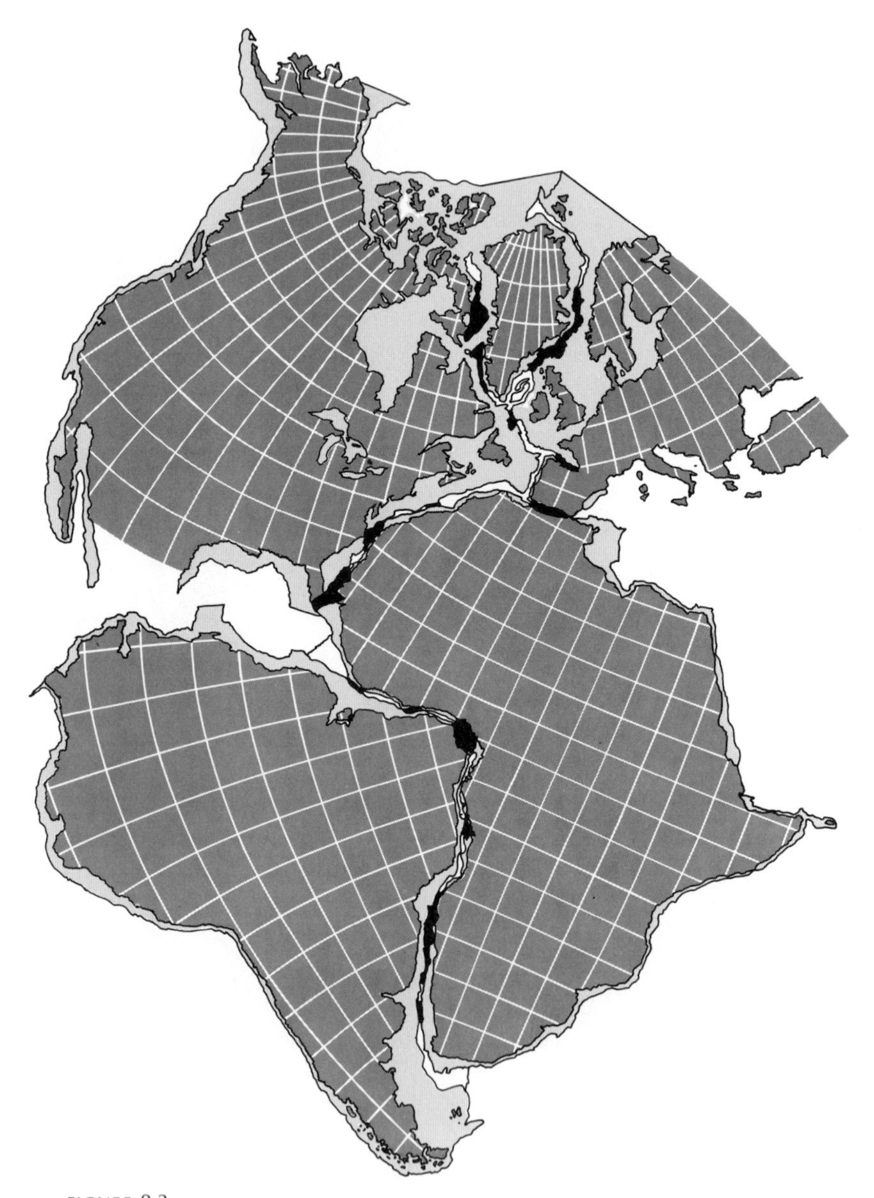

FIGURE 8.3
The fit of the Americas, Europe, and Africa is improved when the edges of the continental shelves are used, instead of the sea-level shapes. Rocks of similar type are then found to be associated, as indicated by the dark areas. (From P. M. Hurley, "The Confirmation of Continental Drift." Copyright © 1968 by Scientific American, Inc. All rights reserved.)

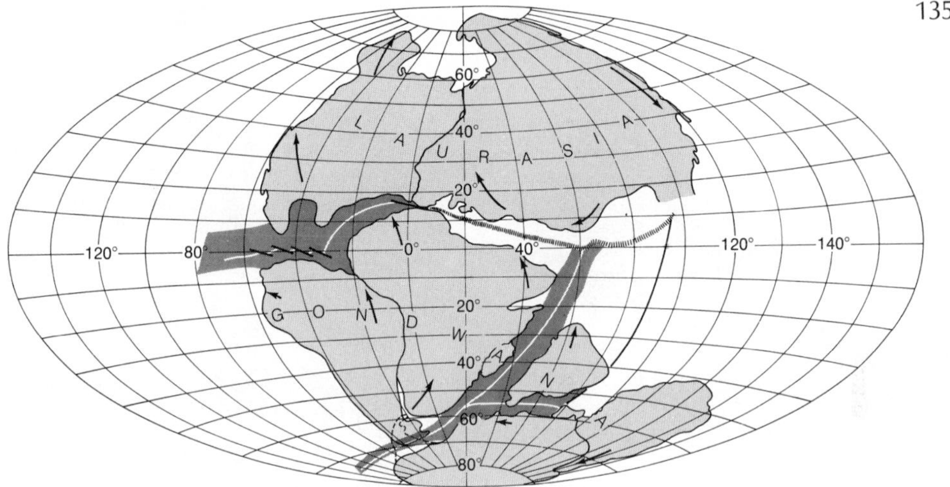

FIGURE 8.4
This association of the continents is believed to have existed about 200 million years ago. (After R. Dietz and J. Holden, "The Breakup of Pangaea." Copyright © 1970 by Scientific American, Inc. All rights reserved.)

Over long periods of time, hundreds of millions of years, the familiar shapes of the Earth's continents undergo major changes. Figure 8.3 shows how the coastline of West Africa matches the eastern coastline of South America. Not only do shapes match, but particular kinds of rock are found to fit when the two coastlines are brought together as they are in Figure 8.3, in which the matching of the rocks is shown by the dark areas.

The explanation given for this situation is simple but remarkable, namely, that Africa and South America were one joined together. Indeed, other sections of the continental land masses were once joined together, in the pattern of Figure 8.4, which shows the continents as they were some two or three hundred million years ago.

The continents are made out of rock that is some 25 per cent less heavy, volume for volume, than the rocks on which the continents lie. The continental rocks are also less heavy than the rocks of the ocean floors. This causes the continents to "ride high." The lighter rock floats in the heavier rock. Figure 8.5 shows how the western United States rides above the floor of the Pacific Ocean.

© 1969 NATIONAL GEOGRAPHIC SOCIETY
KORYAK RANGE
5417
CONTINENTAL SHELF
NUNIVAK I.
1505
1675
−240
−90
−385
KAMCHATKA
PENINSULA
15584
11339
7073
2992
BERING SEA
PRIBILOF ISLANDS
−180
BERING ABYSSAL PLAIN
−13000
−1400
−110
−200
−5700
(10300)
−1700
−390
(15610)
BOWERS BANK
ALEUTIAN ISLANDS
3125
5925
6920
9387
ALEUTIAN TRENCH
−21500
−25200
−23300
−19700
ALASKA PENINSULA
Mt. McKinley
20320
ALASKA RANGE
11258
16390
4234
Anchorage
19850
GULF OF ALASKA
15300
Juneau
COAST MOUNTAINS
ROCKY MOUNTAINS
9500
10002
4100
13104
PRATT SEAMOUNT
−2300
WELKER GUYOT
PATTON SEAMOUNT
−3700
(12300)
PARKER SEAMOUNT
BOWIE SEAMOUNT
Vancouver
Seattle
Mt. Rainier
14410
CASCADE RANGE
CONTINENTAL SHELF
COBB SEAMOUNT
−108
(15892)
12662
13785
9720
14162
3528
−14800
−12100
−16400
ALEUTIAN ABYSSAL PLAIN
−18100
CASCADIA CHANNEL
−17500
−15500
−12100
MENDOCINO FRACTURE ZONE
−16400
−19500
−19200
San Francisco
Mt. Whitney
14494
Death Valley
−282
Los Angeles
6140
San Diego
10154
GULF OF CALIFORNIA
EMPEROR SEAMOUNT CHAIN
−28000
−14800
−18500
TENCHI SEAMOUNT
−21900
−5000
(11000)
SUIKO SEAMOUNT
−18000
CHINOOK FRACTURE ZONE
−19000
−3100
NINTOKU SEAMOUNT
OJIN SEAMOUNT
−17000
KINMEI SEAMOUNT
SHATSKY RISE
−10000
(6000)
−7600
MILWAUKEE SEAMOUNT
−60
(15940)
−18000
−20600
−1140
−14800
ERBEN GUYOT
FIEBERLING GUYOT
−1900
JASPER SEAMOUNT
MURRAY FRACTURE ZONE
−5000
−18400
−19700
−5800
−19800
−2100
MIDWAY ISLANDS
12
35
HAWAIIAN ISLANDS
−15100
−16500
−16100
−16400
−1280
HENDERSON SEAMOUNT
MOLOKAI FRACTURE ZONE
277
−15400
−18200
NECKER RIDGE
HAWAIIAN RIDGE
−19680
Mauna Kea
13796
HORIZON GUYOT
21
(16021)
WAKE ISLAND
HESS GUYOT
CAPE JOHNSON GUYOT
−18600
JOHNSTON ISLAND
−18500
−16500
−1560
(14440)
ISLAS REVILLA GIGEDO
SHIMADA BANK
CLARION FRACTURE ZONE
−15800
−17200
MARSHALL ISLANDS
−17500
−740
CLIPPERTO
−20000

FIGURE 8.5
A part of the Pacific Ocean floor.
(Courtesy of the National Geographic Society.
Copyright © 1968.)

Although the evidence of Figure 8.3 has been available for a long time, scientists were reluctant to believe the obvious implication—that Africa and South America were once joined together—because enormous forces would be needed to tear two such land masses apart. Nowadays the forces are thought to arise from the operation of a kind of nuclear engine within the Earth itself. The energy source for the engine comes largely from the element uranium, which is used to supply the energy output of a man-made nuclear reactor.

What has happened in the Earth is that so much heat has by now been released that an engine has been set in operation,* an engine in which the outer parts of the Earth have been set in motion. The motion is rather complicated: the crust is divided into a number of pieces, "plates," as they are called, which all move with respect to each other. At some places the rocks of a plate emerge from the Earth's interior, as they do along the Mid-Atlantic ridge shown in Figure 8.6. At other places the rocks of one plate may dip below another plate, or below the rocks of a continent.

The continent of Africa rides on a plate, as does the connecting part of Asia. These two plates are different, and their motions are causing them to press against each other. It is this pressing together of plates which has caused a vast line of mountains to be pushed up, starting with the Alps in the west, passing through Turkey and the Caucasus, and eventually reaching Afghanistan and the Himalayas.

*If the outer part of the Earth were initially of a nonuniform composition, with heavy materials like iron mixed with less-dense rock, the release of heat in the outer part would eventually lead to a plastic condition in which the heavy materials would sink gradually toward the Earth's center. This would be one way in which forces adequate to explain the motion of the Earth's surface regions might be generated.

MID-OCEAN CANYON
-10500
-9510
ROCKALL
-1749 (14260)
-780
BRITISH ISLES
London
-306 (15694)
-462 (15538)
ISLAND OF NEWFOUNDLAND
St. Lawrence
-14760
-168 FLEMISH CAP
-15420
-42
CAPE BRETON ISLAND
MIQUELON ISLANDS
GRAND BANKS OF NEWFOUNDLAND
-330
SABLE ISLAND
-228
Boston
-12
LAURENTIAN CONE
New York
-15900
-4500 (11500)
RIFT VALLEY
BISCAY ABYSSAL PLAIN
-15400
-17280
-17700
-15300
KELVIN SEAMOUNT
-5412 (10588)
-4590
SOHM ABYSSAL PLAIN
Washington
MOUNTAINS
IBERIAN PENINSULA
Lisbon
Pico 7615 (23615)
AZORES
-17400
OCEANOGRAPHER FRACTURE ZONE
MID-ATLANTIC RIDGE
CORNER SEAMOUNTS
-16680
Mitchell
HUDSON CANYON
-8520
-4500
-16200
AMPERE SEAMOUNT
-132 (15868)
-870
-24
BERMUDA ISLANDS
259
-9342
-12000
-960 (15040)
6106
MADEIRA ISLANDS
Jebel Toubkal 13665
ATLAS MOUNTAINS
-17400
-18300
-17820
-15000
OUTER RIDGE
HATTERAS ABYSSAL PLAIN
BLAKE PLATEAU
-3900 (12100)
BERMUDA RISE
-12000
-1470 (14530)
-12600
CANARY ISLANDS
12198 (28198)
-13200
Miami
BAHAMA ISLANDS
-18150
NARES ABYSSAL PLAIN
-19200
200
CUBA
-18624
Atlantic Ocean's deepest point
PUERTO RICO TRENCH
-27500
10417 (26417)
-6978 (9022)
-16200
-13500
-12844
-17060
GREATER ANTILLES
CAPE VERDE ISLANDS
CARIBBEAN SEA
LESSER ANTILLES
KRYLOV SEAMOUNT
-4260
-17753
9281 (25281)
Dakar
Niger
-12926
-16594
GAMBIA ABYSSAL PLAIN
-15748
-6135
-15994
-8333

FIGURE 8.6
A part of the Atlantic Ocean floor.
(Courtesy of the National Geographic Society.
Copyright © 1968.)

Sometimes a plate presses against a continental mass, as it is doing along the western coastline of the Americas, causing the mountain ranges of the Rockies and Andes to be formed. In still other places, the rocks of two plates dip down together, producing the deeps of the ocean floor. The whole complex of all these plates, over the whole surface of the Earth, is shown in Figure 8.7. It is their motion that causes volcanoes and earthquakes: volcanoes result from the heat generated by friction when rocks grind against each other, and earthquakes result from the actual motions themselves. These dramatic phenomena occur when one plate abuts another. In contrast, the central regions of the plates, far removed from their edges, tend to be rather quiescent. This is why some places on the Earth are violent and some peaceful; which one a place will be depends on where it is with respect to the system of plates.

We tend to think of earthquakes and of the outbursts of volcanoes as disasters, and we tend to regard mountain ranges as barren regions. So we might think that the plate motions are not an advantage to human life. Yet on a broader view, we would be quite incorrect. Without this continuous turning over of the Earth's crust, there would probably be no mineral deposits on the Earth's surface. Suppose there were a range of 20,000-foot mountains in the center of Australia. It would bring down vast quantities of snow, and large rivers would flow out of such a mountainous area, which would provide much needed water throughout what is now the great Australian desert. So we can see that, without the system of plate movements, the surface of the Earth would almost certainly be far more inhospitable to man. Indeed, the gradual wearing down of the continents might well smooth out the Earth so much that, instead of

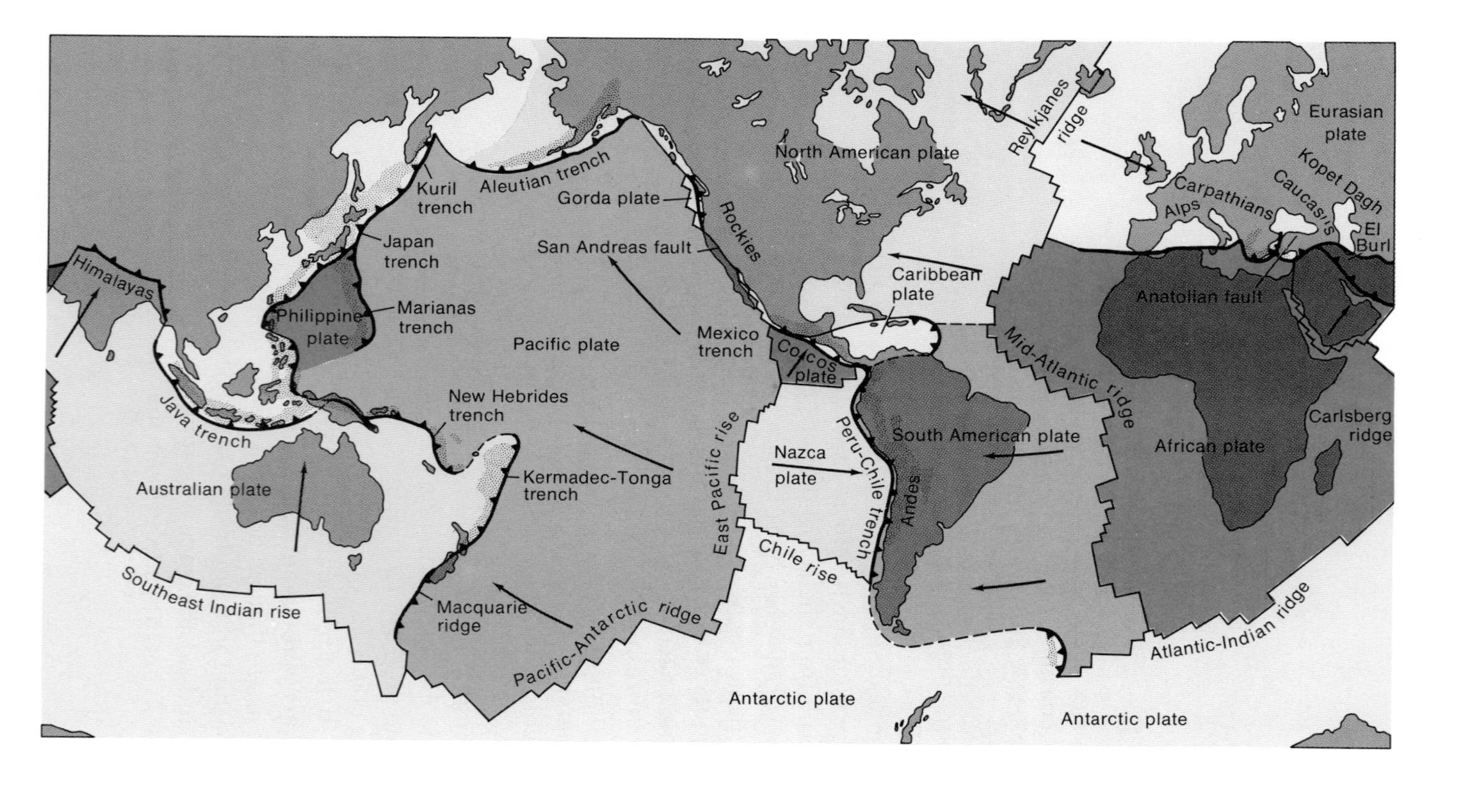

FIGURE 8.7
Showing how the surface of the Earth is broken up into plates.
(From Allen J. Dewey, "Plate Tectonics." Copyright © 1972
by Scientific American, Inc. All rights reserved.)

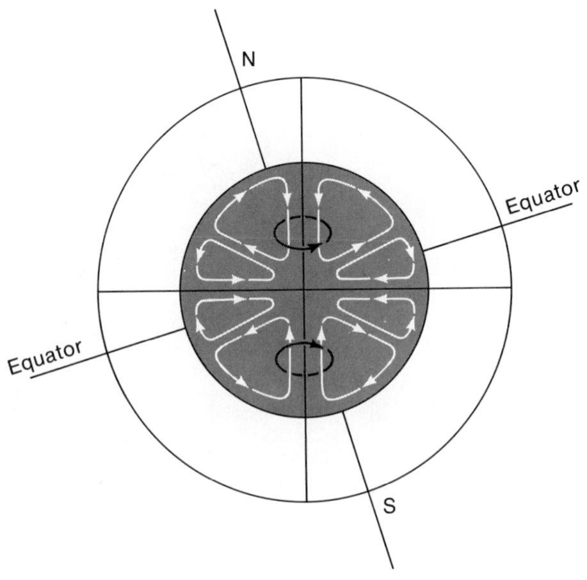

FIGURE 8.8
This kind of circulation, in meridian planes in the Earth's core, could maintain the magnetic field of the Earth.

the continents sticking out of the ocean, water would entirely cover the whole surface of the Earth. The nuclear engine inside the Earth thus does much more than provide us with interesting scenery; it provides the environment for life as we know it.

Earthquake shocks set the whole body of the Earth in vibration. Seismic measurements of these vibrations, together with data from laboratory experiments on the properties of materials under high shock pressures, give strong reasons for believing that the Earth has a metallic core extending to a radius of about 3,400 kilometers. This is rather more than half the total Earth radius of 6,370 kilometers. Outside the core lies a mantle of rock, the surface of which we have just considered in some detail. The core is thought to consist mostly of molten iron at a temperature of about 5,000 degrees. Mixed with the iron are likely to be smaller quantities of the iron-group metals—titanium, vanadium, chromium, manganese, cobalt, and nickel. Electric currents flowing in the metal generate the magnetic field of the Earth, possibly from the kind of liquid motion shown in Figure 8.8.

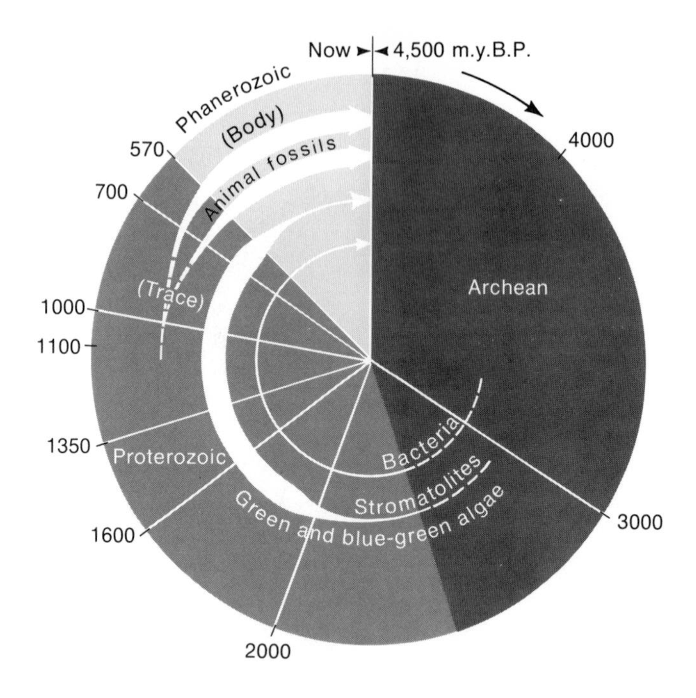

FIGURE 8.9
The early forms of life were bacteria and the blue-green algae.

If we think of a human generation as occupying 45 years, the Earth has been moving year by year around the Sun for as long as 100 million human generations. A way of representing this great span of time is shown in Figure 8.9. Imagine the hand of a clock starting in the 12 o'clock position, and imagine it making a complete circuit, but taking 4,500 million years to do so. That is to say, the hand starts 4,500 million years back into the past, at the time the Earth was formed. The numbers marked on the outer circle of Figure 8.9 refer to moments of time in the past—4,000 means 4,000 million years ago, 3,000 means 3,000 million years ago, and so on. Also shown in Figure 8.9 are the broad forms of life that have existed in the past on the Earth. The life forms with which we are normally familiar, the complex forms, all belong to the last sector, starting at the number 570 (570 million years ago) and continuing through until now.

Life has existed on the Earth for about 3,500 million years, but at first only in the simple form of bacteria. Next came the green and blue-green algae, going back nearly 3,000 million years. These early forms can all live at quite high temperatures. Some forms of bacteria can indeed survive at the temperature of boiling water. The algae grow in profusion in volcanic hot springs like those of Yellowstone National Park. The fact that the early life forms were of this high-temperature kind suggests either that life started in hot springs, or that the whole Earth was very much warmer than it is now.

Let us now go back in time to the starting point of Figure 8.9, to the situation some 4,500 million years ago when the Earth was formed. How did our planet then come into being? A very general answer to this question can be given by associating the origin of the Earth and the other planets of our system with the process in which the Sun itself was born. Two curious facts, one relating to the Sun and the other to the Earth, give clues to what the details of this process may have been.

The early solar condensation must have been rapidly rotating. Yet the present-day Sun spins slowly. The primaeval Sun must therefore have been robbed of its rotation, of its angular momentum, to use the appropriate term. On the other hand, the planets in their orbital motions around the Sun have remarkably large angular momenta. It thus seems likely that the process which robbed the Sun of its initially rapid rotation operated to transfer the associated angular momentum to the planets.

Angular momentum is a quantity that can be neither created nor destroyed; it can only be transferred from one place to another. The transfer to the planets caused them to become quite widely spaced from the Sun. If we think of the present-day Sun as being represented by a ball 6 inches in diameter (about the size of a grapefruit), the inner planets, Mercury (Figure 8.10), Venus (Figure 8.11), Earth, and Mars (Figure 8.12), are at the distances of 7, 13, 18, and 27 yards

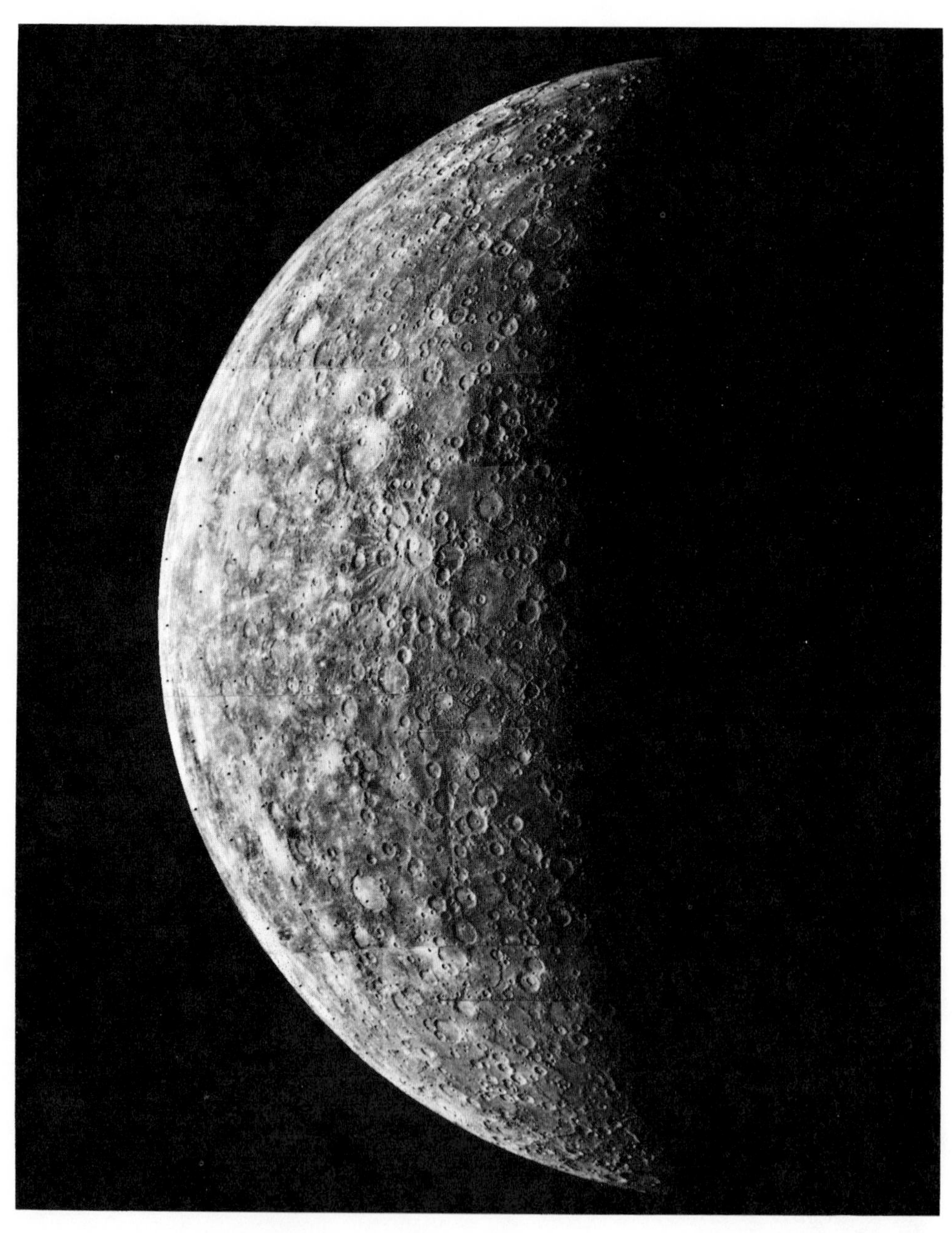

FIGURE 8.10
A recent picture of the planet Mercury. (Courtesy of NASA.)

FIGURE 8.11
Venus photographed from a distance of 700,000 kilometers by ultraviolet television cameras. Winds circulate around the planet in about four days. The clouds may be droplets of sulfuric acid. (Courtesy of NASA.)

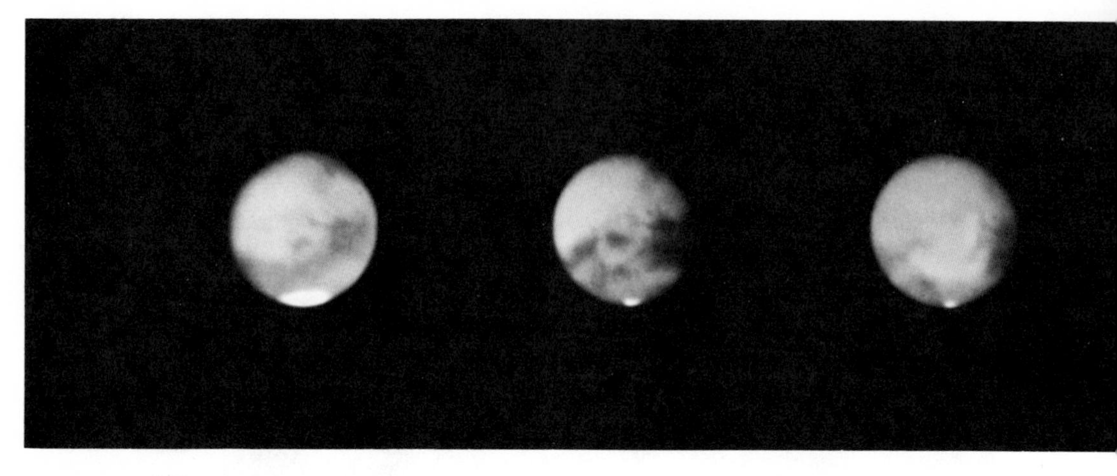

FIGURE 8.12
Ground-based photographs of Mars, showing the effect of a dust storm in the righthand picture. (Courtesy of the Lowell Observatory.) Pioneer II:

from the Sun, respectively, each one being not more than the size of a pin's head. The great planets, Jupiter (Figure 8.13), Saturn (Figure 8.14), Uranus (Figure 8.15), and Neptune (Figure 8.16) are of the sizes of small peas about 90, 170, 350, and 540 yards, respectively, from the Sun. The transfer of angular momentum, by whatever process it occurred, would necessarily have pushed the gases from which the planets eventually formed out to these comparatively large distances from the Sun.

The second clue to the process by which the planets were formed comes from the selection of the elements that went to form the Earth. Elements which are gaseous, or which form gaseous compounds, at temperatures of a few hundred degrees are largely absent from the Earth. Elements and compounds that form highly refractory solids are apparently present in the Earth in essentially their cosmic proportions. A selection process operated therefore to separate gases from solids, and it operated at temperatures at least as high as several hundred degrees, perhaps as high as 1,000 degrees.

FIGURE 8.13
A view of Jupiter that cannot be seen at all from Earth, showing convection cells near Jupiter's north pole. These cells rise like thunderstorms do on Earth. (Courtesy of NASA.)

FIGURE 8.14
The planet Saturn as seen in early 1973. The prominent dark gap in the ring is known as "Cassini's division."
(Courtesy of New Mexico State University.)

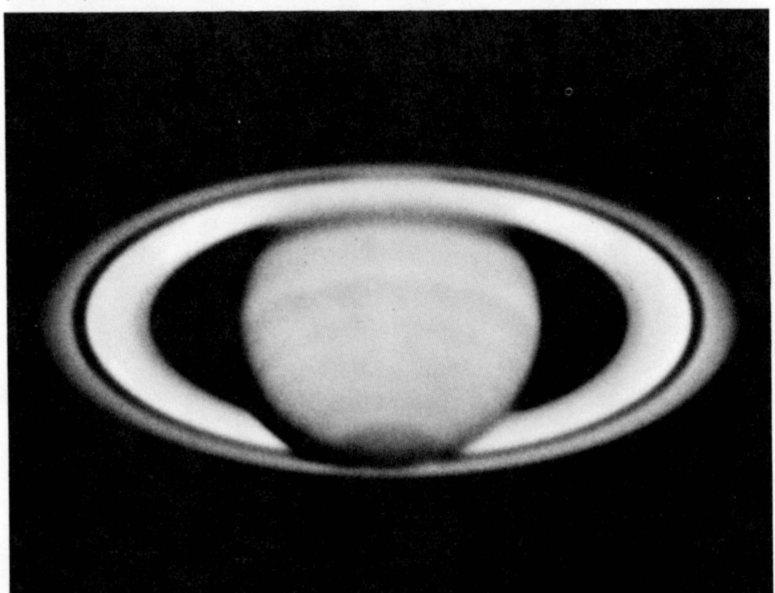

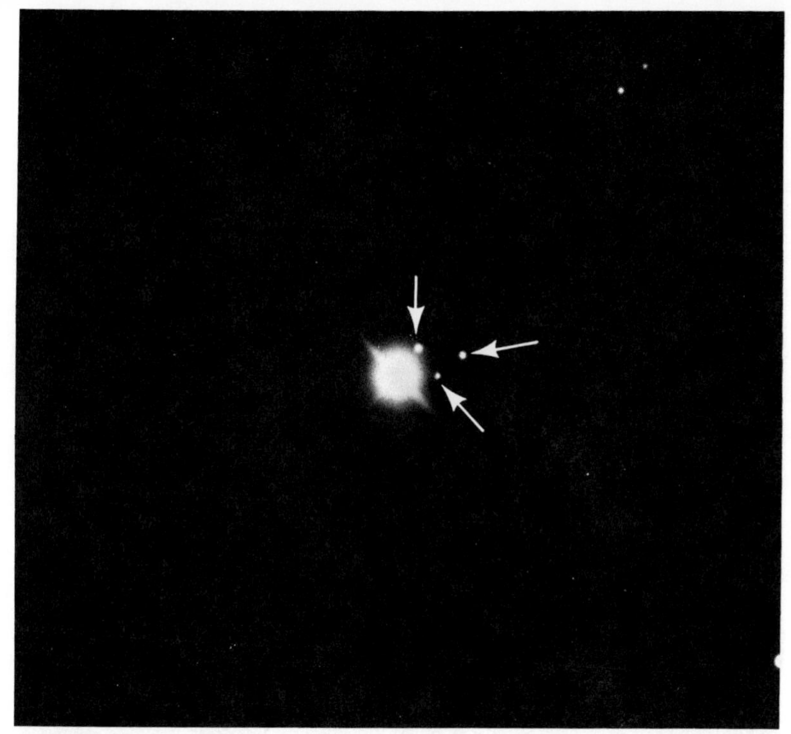

FIGURE 8.15
The planet Uranus and three of its satellites.
(Courtesy of the Lick Observatory.)

FIGURE 8.16
The planet Neptune and the satellite Triton.
(Courtesy of the Lick Observatory.)

Combining the two clues permits us to say that, as the planetary gases moved outward from the Sun, certain elements and compounds condensed into solid particles. Since aggregations of solid particles would be left behind as the gases continued their outward motion, a separation of gases and solids occurred quite naturally, with the inner planets, Mercury, Venus, Earth, and Mars, forming from the solid aggregations, and with the great planets, Jupiter, Saturn, Uranus, and Neptune, forming at a later stage from the gases which continued their motion to the outside of the solar system.

Most single stars of comparatively small mass spin slowly, as the Sun does. The reasoning followed above, leading to the formation of a planetary system, is applicable to all such stars, of which there are about 10^{11} in our galaxy. There are on the order of 10^{9} galaxies visible in a large telescope. With 10^{11} planetary systems forming in each galaxy, we have, within the range of our telescopes, some 10^{20} possible abodes of life.

A year or so ago I set off one winter morning to walk in the mountains of the English Lake District, which is where I live. The top ridges of the hills are rough and rocky, with steep slopes for such small mountains—the whole area is an uplifted dome that was severely cut about by the glaciers of the last ice age, about 10,000 years ago. As I climbed a straightforward track up into the higher regions, I knew from ominous cloud formations that a storm was about to break. I decided overoptimistically that I would still have time to complete a two-to-three-mile traverse of the high ridges before the full fury of it was upon me. On a calm summer day this particular traverse is no more than a pleasant walk. In the storm of that winter day it was very much otherwise. The path, running through a maze of ridges and rocky turrets, was obliterated in the snow. The gale threw hail into my eyes with such stinging force that I had to grope forward with them shut for nine parts of the time.

TABLE 8.1
Planetary data

Planet	Half of long axis of orbit (km)	Eccentricity	Average orbital speed (km/sec)	Time around orbit (years)	Mass		Radius (km)	Density (gm/cm³)	Axial rotation period	Tilt of rotation axis to orbital plane
					(gm)	Earth masses				
Mercury	5.791×10^7	0.206	47.90	0.2408	3.3×10^{26}	0.056	2439	5.4	58.7 days	7°
Venus	1.082×10^8	0.007	35.05	0.6152	4.9×10^{27}	0.81	6050	5.1	243 days	6°
Earth	1.496×10^8	0.017	29.80	1.0000(4)	6.0×10^{27}	1.00	6378	5.52	$23^h\ 56^m$	23.5°
Mars	2.279×10^8	0.093	24.14	1.8809	6.4×10^{26}	0.11	3394	3.97	$24^h\ 37^m$	24°
Jupiter	7.783×10^8	0.048	13.06	11.8622	1.9×10^{30}	318	71,880	1.33	$9^h\ 55^m$[a]	3°
Saturn	1.427×10^9	0.056	9.65	29.4577	5.7×10^{29}	95	60,400	0.68	$10^h\ 38^m$[a]	27°
Uranus	2.869×10^9	0.047	6.80	84.013	8.8×10^{28}	15	23,540	1.60	$10^h\ 49^m$	98°
Neptune	4.498×10^9	0.008	5.43	164.79	1.0×10^{29}	17	24,600	1.6	15^h	29°
Pluto[b]	5.900×10^9	0.249	4.74	248.4	—	—	—	—	—	—

[a]Temporate zones
[b]Data for Pluto for last six columns is uncertain.

The summit rocks, encased in ice, were treacherous for a lone man, and I was forced to avoid them by a long detour. I was glad at last to be able to find my way down through a steep rocky cleft. There are many such clefts on that hillside. If you are lucky enough to find the right one, you come out on a still steep but pleasant grassy slope—otherwise you must stumble downhill for several thousand feet over awkward crags and piles of stones. I was lucky. I came out onto the grass slope, and immediately saw there a mountain sheep placidly munching tufts of grass where the snow had been swept thin by the wind.

That evening, back at home, I sat before a blazing log fire, listening to music. Oddly enough, I had no taste for my favorite composers. It was to Sibelius I turned, to the second symphony with its powerful slow movement. As I sat listening, I was at last aware that my day, starting so innocently in the morning, had found a quality quite unique in a whole lifetime of walking among mountains. Emotionally, I had experienced an intense conviction that I was seeing the Earth the way it was far far back in the past, some 4 billion years ago. The gale, the snow, the rock, could all have been the same, soon after the Earth itself was formed. In my half-blind trudge, alone on the high ridge, and in the failing light of a short winter day, the mountains seem to go on and on forever—it seemed impossible there could be any other moment of time in that white-shrouded world of the gale. Then suddenly I sighted the sheep, and in that moment passed from a stark, primitive world back to the world of life. Even the few ragged blades of grass, projecting through the snow, had seemed more luxuriant than a spring garden filled with blossom trees and flowers. As an economist might have seen it, the sheep was worth no more than $30, but to me the sheep was evidence of the miracle of life on the Earth, and in this sense it was beyond price.

FURTHER READINGS

Continents Adrift: Readings from Scientific American. W. H. Freeman and Company, 1972.

F. Press and R. Siever, *Earth.* W. H. Freeman and Company, 1972.

9

THE BIOLOGIST'S UNIVERSE

CHARLES ROBERT DARWIN
(1809–1882)

(Photo from The Bettmann Archive.)

9

THE BIOLOGIST'S UNIVERSE

Understanding how the inanimate ancient world was transformed into an Earth teeming with life involves a hard and subtle problem—although the story seems to have begun without undue difficulty, indeed, more simply than would have seemed possible only a few years ago. It is widely agreed that life was based on a few simple molecules, particularly water (H_2O), hydrogen cyanide (HCN), carbon dioxide (CO_2), and ammonia (NH_3). These molecules have recently been found to exist in vast quantities within the gas clouds of the Milky Way, like those shown in Figure 9.1. The ingredients of life are evidently available in ample supply.

What the first step was toward the origin of life, according to ideas proposed by A. I. Oparin, is also comparatively well-understood. This is a step in which these simple inorganic molecules were built into more complex ones containing perhaps 10 to 30 atoms,

FIGURE 9.1
The Trifid Nebula. (Courtesy of the Hale Observatories.)

organic molecules like the amino acids, twenty of which play a critical role in building the much larger structures that are important in living material. The essential feature of this first step is that it supplied a store of energy which could then be used to drive more complex systems. The source of the energy must have been radiation received from the Sun. This first step does not seem unlikely—indeed, most investigators seem to think it a highly probable event. This being so for the Earth, we may take similar events to have happened in very many, if not all, of the 10^{20} conceivable abodes of life.

So far, so good. Yet with such an energy store, we are still far from a synthesis of the large molecules on which life itself is based. Life as we know it depends on chemical reactions involving molecules called proteins. Although our diet requires a daily intake of proteins in our food, it is important to realize that what we need is not the ingested proteins themselves, but the amino acids that make up these proteins. What happens in our bodies is that the ingested proteins are first broken down, and the resulting amino acids are then rebuilt into our own proteins, the proteins required specifically by our kind of creature, the human. A dog proceeds similarly, using the same basic amino acids, but building them into protein structures suited to itself. And so for all the animals. We all use the same twenty amino acids, but we arrange them individually according to our separate needs.

How is this done? How does each animal manage to build just what is right for itself? Nowadays we even understand the answer to this crucial question. Each of us contains a kind of vast chemical blueprint, which is copied time and time again, as our proteins are made to serve our various bodily functions. Here emerges the hard and subtle problem. We can understand, from major recent achievements in microbiology, how the chemical reactions on which life is based are controlled by complex molecules like the proteins. We understand how by a copying process these complex molecules are built up from simpler molecules like the amino acids. But how did the simpler molecules get together to form the complex ones in the first place? For not until the first complex molecules were formed could a blueprint for copying them be constructed. Once it was started, we can see how the system worked, but how was it started?

We would have a straightforward answer to this question if we could reasonably argue that the initial arrangement of a living cell was due to chance. But a probability calculation soon shows that such an explanation is not reasonable, that we are faced with a strange situation. Even the least complex protein that is biologically important is made up from about a hundred amino acids linked to form a long chain. Each link of the chain consists of one particular amino acid taken from a set of twenty. Yet with twenty possibilities for each link, there are $(20)^{100}$ possible distinct proteins that have a hundred links. How was the particular arrangement that is biologically important picked out from this enormous number of possibilities? In a suitable environment, many arrangements could certainly be tried, but not remotely as many as $(20)^{100}$. The total number of amino-acid molecules on the whole Earth cannot have been much more than 10^{44}. If each one were involved every few seconds in a new trial, even four billion years would not be time for more than about 10^{60} trials, vastly less than the $(20)^{100}$ trials needed. Even for all the 10^{20} possible planetary systems in the observable universe, there could only be about 10^{80} trials, still much less than $(20)^{100}$. Plainly then, most of the ways in which amino acids might be linked to form proteins can never have been tried, not even once in the history of the whole of the visible universe. What then, we may ask, were the criteria of judgment used to decide which arrangements would be tried and which not?

There appear to be two criteria for such selection. Imagine a protein chain, partially completed, to which a further amino-acid link is to be added. In general, not all the twenty amino acids will be available in equal concentrations. Those present in high concentration will, obviously, be more likely to form the next link than those in low concentration. And even if the concentrations were all equal, some amino acids would still be more likely to attach themselves than others—because of the physical relationship between the still-incomplete protein chain and the different structural details of the twenty amino acids. For both these reasons, some arrangements would be more likely than others. The answer to our question lies therefore both in the environment and in the basic physical laws themselves. These two factors cause some arrangements to be favored, others to be discriminated against.

Some years ago, J. D. Bernal, in his book *The Physical Basis of Life,* remarked on the peculiar affinity that clay surfaces have for biologically important molecules. Clay surfaces could well have provided the environment for the building of trial protein structures, in which case the properties of the clay itself would affect the concentrations of the different amino acids, and would hence have provided an important selective criterion, causing some structures to be tried and others not.

Let us now look at the strange situation mentioned above. The critical question is, *why* did the biologically relevant proteins (and some other large molecules) happen to be among the very small fraction of possible arrangements that were selected by these criteria? One answer to this question might be that many alternative systems of proteins could have produced life equally well. Then there would be no need for the amino acids to have been shuffled into a more or less unique system; any one of these alternatives would have sufficed. In short, life could have happened in very many ways. A consequence of this point of view is that life would be widespread throughout the universe.

If, on the other hand, life does depend on the generation of a more or less unique system of large, biologically important molecules, then we must look for an interesting connection between the environmental situation, the physical laws, and the origin of life itself. Rather than a mere proliferation of possible origins, we would then have a compact association of physical laws, environment, and life, an association that would bring together the disciplines of the physicist, the astronomer, the geophysicist, and the biochemist. My own preference is for this second answer; so, for the purposes of the following discussion, I will take it to be correct. Let us see where it leads.

First, it also leads to the conclusion that life is likely to be widespread throughout the universe, but for a reason different from the mere proliferation of possibilities. Life must be widespread because the same physical laws and similar environmental situations apply for all the 10^{20} possible abodes of life that lie in the 10^9 galaxies visible with a large telescope. The crucial step of selecting the biologically important large molecules has a yes-or-no character. The fact that the Earth was a "yes" may be taken to imply a "yes" in many other places.

All these developments were completed on the Earth during the first 1,000 million years, as can be seen from Figure 8.9. The fossil evidence for the existence of bacteria reaches back some 3,500 million years into the past—and this is in no sense a limit. There could have been life even earlier.

The early life forms shown in Figure 8.9—bacteria, green and blue-green algae—can all survive at high temperatures. This fact suggests, I believe quite strongly, that life on the Earth has followed a sequence of declining temperature. This might have happened if life originated in hot volcanic pools, then spread into cooler waters, or it may have happened because the Sun has become less and less bright as time has gone on. If the latter is true, we can begin to understand why it took so long before complex life forms appeared on the Earth. The need during most of the Earth's history to withstand a very high temperature must have imposed constraints on

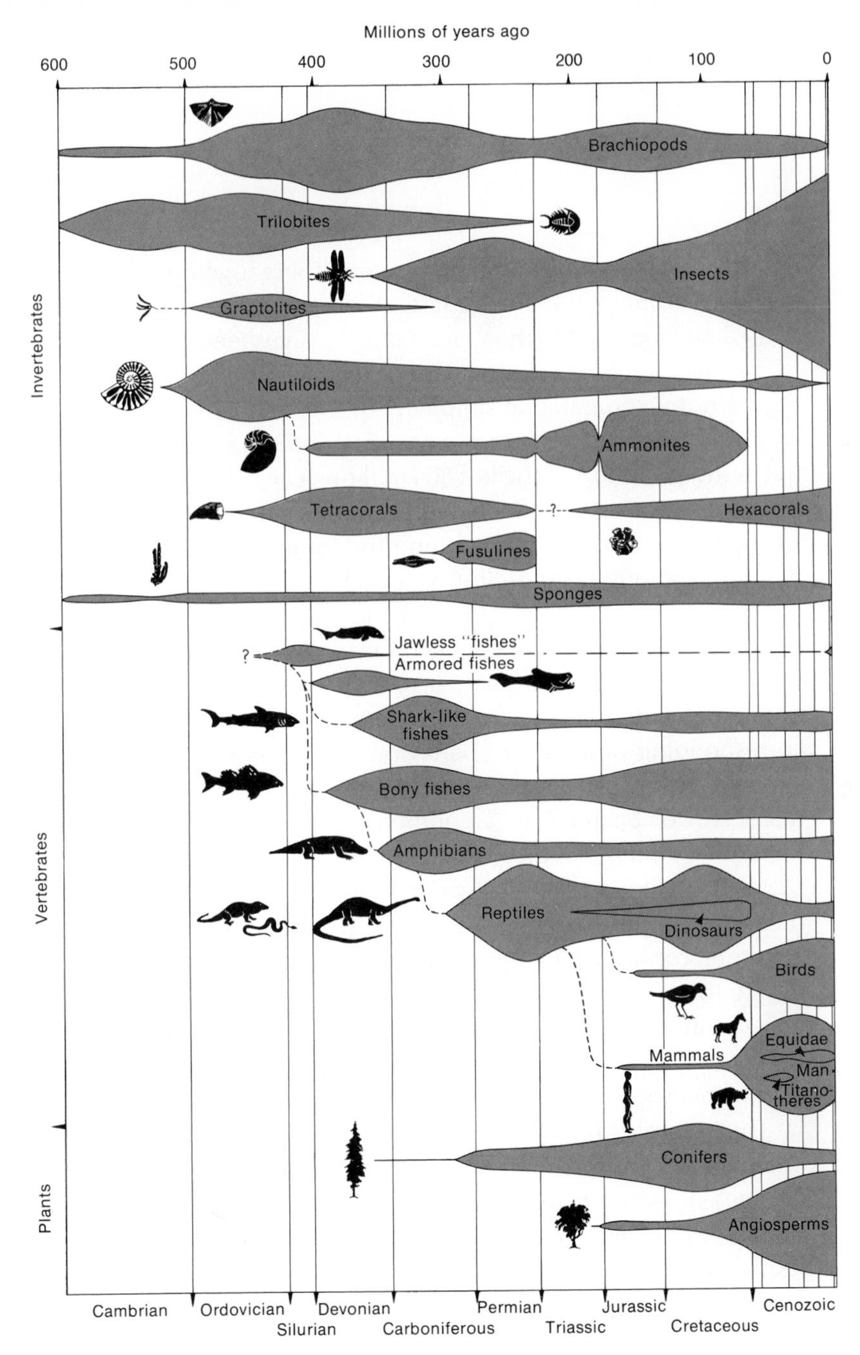

FIGURE 9.2
How evolution is believed to have occurred on the Earth. Geological ages are given along the bottom. Notice the connection of one sequence of evolution with another, denoted by dotted lines. (From Shelton, *Geology Illustrated.* W. H. Freeman and Company. Copyright © 1966.)

the cells, perhaps inhibiting them from joining together into multi-celled creatures. Only in the last 1,000 million years has the Sun cooled sufficiently for this constraint to have been removed.

In such a way it seems possible to understand the difference between the comparative simplicity of the early life forms and the complexity of the forms that have arisen more recently. Figure 9.2 traces the evolution of the last 500 million years. The richness of this evolution contrasts markedly with the stark simplicity of the forms which dominated the Earth during the long time span from 3,500 million to about 1,000 million years ago.

The changes of Figure 9.2 are generally believed to have been directed by a further selective process. There are always slight variations from one member of a species to another in the structure of the molecules which carry the genetic makeup, and these cause corresponding variations of behavior: one member may be slightly stronger than other members, or slightly more resistant to a particular strain of bacteria. In a competitive situation, such a mutant member would be favored, would be more likely to grow to maturity, and would be likely to generate a larger number of surviving offspring. If, furthermore, the original variation of the genetic structure continued in the offspring, the same advantage would pass to the offspring, and hence would in turn influence the second generation in the same way. The original variation would thus come to spread itself more widely through the species, and ultimately the whole species would come to be endowed with it.

This interplay between initially random variations and an evolutionary selection is thought to have controlled and directed the changes of Figure 9.2. But could such a process produce the exceedingly rich aggregate of plants and animals that has emerged on the Earth during the past 500 million years? Charles Darwin answered this question affirmatively in his book *The Origin of Species*. By noting variety changes in birds and plants, Darwin sought to estimate the dynamic rate of the evolutionary process. His conclusion was that the rate was sufficient to explain the changes of Figure 9.2. Although this analysis constituted a crucial advance of biological thinking in the mid-nineteenth century, it is scarcely adequate to meet our twentieth-century curiosity. Indeed, I think today we can even feel some sympathy for Darwin's opponents in the classic debate which followed the publication of *The Origin of Species.* The counterargument, phrased in modern terms, starts by noting the enormous quantity of information required to specify the structures of plants and animals. What was the source of this information? Darwin's researches showed how the information might be shuffled about, but the source of the information was not identified. The implication of nineteenth-century biology seemed to be that the information had somehow been generated spontaneously, essentially out of nothing. This view is unlikely to be correct, however, because the chance that many individual cells could shuffle themselves by a random process into a complex life form is exceedingly small.

Simple beginnings can be ascribed to chance, but it is the basic physical laws which then decide that the simple beginnings can serve as the building blocks of more complex structures. And it is also the physical laws which further decide that the structures so produced have remarkable new properties *not apparent in the beginning.* The physical laws contain within themselves a sequence of mounting levels of interest—inorganic molecules building simple organic molecules, simple organic molecules building complex ones, complex long-chain organic molecules forming into a biological cell able to replicate itself, many such cells building into multicelled creatures, multicelled creatures interacting with each other . . . , and so on.

The remarkable nature of this sequence is well-illustrated by a comparison with purely inorganic forms. There is no shortage of inorganic structures containing quite large numbers of atoms. A single crystal of rock, or of a mineral, or of a snowflake, contains subtleties of great interest. But such crystals do not fit together into larger patterns of still greater interest and complexity. Large quantities of inorganic material only *repeat* the simpler forms. The information content of a huge quantity of rock is not much greater than that of a single stone. The information content of a blizzard is essentially the same as that of a single snowflake. There is no hierarchy of structure, with one level of subtlety piled on another. It is the existence of such a hierarchy which characterizes biological systems. Each level in the hierarchy serves as the building block for the next level, apparently in an ever-expanding sequence. The fact that it can do so is determined by the physical laws, which therefore seem to contain within themselves the information we have been seeking.

The above considerations become immediately relevant as soon as we seek to broaden the picture of Figure 9.2. As we noted at the beginning of this chapter, the basic life-forming molecules are widely diffused everywhere in our galaxy and in other galaxies too. It is therefore natural to suppose that life is likely to be widespread throughout the universe. Structures similar to ourselves can be expected, simply because of the vast profusion of planets and stars. There are 100,000 million stars in our galaxy alone, and probably a considerable fraction of them have planetary systems. Supposing this expectation to be correct, consider what Figure 9.2 might be like in these other systems. Combining the different forms of Figure 9.2 for all such systems, we arrive at a kind of galactic zoo. What, we might ask, are the variations of living forms in this zoo likely to be? One answer might be that the variations of form are enormous, stretching beyond the wildest flights of our imagination. An opposite answer would be that Figure 9.2 taken for our own planet alone already contains a fair representation of the full range of possibilities. Let us see if we can develop criteria for making a judgment between these view points.

The same physical laws apply in other places, and so should have permitted the building of a biochemical and biological hierarchy in much the same way as in our own solar system. The existence of our particular kind of hierarchy is so remarkable in itself that it must surely represent a mainstream for the whole universe. Hence it seems likely that very many evolutionary systems will bear a general resemblance to Figure 9.2.

Environmental factors may be expected to lead to both startling similarities and marked differences. Let us consider the similarities first, especially as they apply to the emergence of animals. We distinguish between higher and lower animals according to the complexities of the nervous systems with which they are endowed. A nervous system is basically electrical in its operation, with an animal made up of a chemical system plus an electrical one,

$$\text{Animal} \equiv \text{Chemical replication} + \text{Electrical system.}$$

The more the electrical part dominates this summation, the "higher" we judge the animal to be in the evolutionary scale. The more the electronic system happens to match our own system, the more highly we regard the animal. At a certain level of electronic complexity, a little below our own, we rather arbitrarily introduce the notion of "intelligence." Essentially as a matter of definition, any creature with an electrical system more complex than our own would be endowed with "high" intelligence.

Animals are not readily able to synthesize amino acids and sugars, as plants do. Animals must therefore acquire these substances, either by eating plants or by eating other animals. Basically, all animals are scroungers, living on the stored chemical potential which others have first accumulated. It was precisely to assist in the process of scrounging that the electrical systems possessed by animals developed. Since the better the electrical system, the better the scrounger, biological evolution on the Earth has operated steadily for many millions of years to increase the complexity of animal electronics. And since we judge the level of an animal by this complexity, it follows that the higher the animal, the greater the scrounger—with Man himself sitting at the top of the pyramid. The electrical system in Man has indeed become so subtle that our scrounging, for energy in particular, has now extended well beyond the eating of plants and of other animals. We scrounge extensively today on nonliving materials. The discovery of fire made use of the decay products of trees as an energy source. The burning of coal and oil were further steps along the same path; and in the modern nuclear power plant we use entirely nonorganic materials as an energy source.

We can expect similar developments to have occurred in many places, perhaps in many millions of places. It seems clear that the development of an electrical system would be very likely for all animals, everywhere. Because of the need to search for food, "eyes" would be a normal development. Animals with eyes are then likely to prey on each other, with evolution forcing the development of "weapon systems"—claws, teeth, and, ultimately, the most deadly weapon of all, a complex brain. The logical sequence leading to the emergence of a thinking brain (i.e., with consciousness) appears inevitable, and we can expect it to have happened quite generally.

So much for the similarities between our own and other cases. What of the differences? Light from the Sun, when analyzed with respect to wavelength, has maximum intensity in the blue-green part of the spectrum. It is surely no accident that our own eyes are most sensitive in the same general range. The half-hour in the evening before dark must always have been critical to survival, critical to predators and to their prey alike. Animals equipped with eyes more sensitive to light than those of their enemies, or those of their prey, would evidently be better fitted to survive than their competitors. Mutations favoring increased light sensitivity have therefore always been highly advantageous, and have operated to make our eyes most effective at just the wavelengths where there is most light, the blue-green color range.

A similar selective process would be likely to occur in other places, but because the central star might emit light whose maximum intensity falls in a rather different color range, the process would operate to produce eyes whose greatest sensitivity falls in that other color range. It is in this kind of respect that we may expect other creatures to be different from ourselves. The creatures would not possess one eye or three eyes, because two eyes are geometrically more advantageous than one or three. They would differ in the makeup of the retina of the eye itself, and perhaps in the exterior pigment, say, red, instead of our terrestrial brown, black, and blue eyes.

If gravity were less than it is on the Earth, or if the air density were greater, there is no doubt that flying animals would be better placed in the evolutionary process. As things are on the Earth, birds are nearly too heavy to fly at all. They do so only by paring their weight to a minimum: they have hollow bones and brains of minimal capacity, and for long-ranging birds even the ability to come freely down from the air onto the land is denied. The swift, for example, must always nest in such a position that it can take to the air simply by falling out of the nest. If gravity were less, it would be possible for far-flying birds to be equipped with more armament for attacking land-based creatures. Instead of carrying off a lamb, an eagle might carry off a man. Still more formidable, it would be possible for birds to be equipped with thinking brains. In such circumstances, it would be unlikely that a land-based creature like Man would become established at all.

The implication of these considerations is that, although the full-scale zoo, comprising details like those in Figure 9.2 for all creatures on all planets, would be wider in many interesting respects than Figure 9.2 itself, the differences would lie in details, not in principle. The basic driving laws would be the same everywhere, and the environment would not be all that different from one planetary system to another. We have remarked several times that there may well be on the order of 10^{20} planetary systems, with about 10^{11} of them in each of the galaxies (some 10^9) that are visible with a large telescope. Although life would probably not arise in every such system—the chance of there being a planet like the Earth, at the appropriate distance from an appropriate central star, with a similar rotation speed and similar chemistry, might be perhaps one in a hundred—the total number of systems with life would nevertheless be very large indeed. Evidently then, the whole system of life, the aggregation of all diagrams like Figure 9.2, must contain very much redundancy, very much repetitious overlap from one diagram to another. What, one feels inclined to ask, is the reason for this redundancy?

The asking of this question implies that life, taken on a universal scale, serves a purpose—that a reason for it exists. We do not know this to be true; so the question itself may be improper. Yet the fact that life exists at all implies that the basic laws have some remarkable properties. It is hard to see why the basic laws should have these properties unless there is a "purpose" in the whole scheme of things. By "purpose" I mean a suitable linkage between the laws and their consequences, *as if* the laws were aware of their consequences. Certainly the consequences are aware of the laws—we humans are consequences, and we are indeed aware to some degree of the laws themselves. So the linkages appear to go both ways:

$$\text{Laws} \rightleftarrows \text{Consequences.}$$

So much is clear. What is not clear is whether the laws are *indifferent* to their consequences, as the physicist and mathematician usually suppose them to be. But since indifference implies lack of interest, it does not seem reasonable that indifference could lead to such a remarkable looped relation between the laws and their consequences. Hence we may feel that our question is worth pursuing further, even though in asking it we have moved onto insecure ground.

We saw above that life is an hierarchical structure. At each step upward in the hierarchy, possibilities are lost. Even before life arises, possibilities are lost—we just noted that perhaps only one system in a hundred contains a planet like the Earth, thereby reducing the number of sites for life from about 10^{20} to about 10^{18}. There would be further reductions at each hierarchical step described earlier in this chapter, the step from simple organic molecules to complex long-chain ones and thence to a biological cell, the step to multi-celled creatures, the step to intelligent life. But the whole tenor of our argument was that each step, although most remarkable in itself, was not all that improbable. The number of populated sites would be reduced to fewer than 10^{18}, but would still be very large, and so there would be a great redundancy in the evolutionary scheme of Figure 9.2.

Suppose now, however, that Figure 9.2 is seriously incomplete. Suppose the evolution of Figure 9.2 is only a beginning; suppose that many further steps in the hierarchy of life remain possible. There will then be further reductions in the number of cases which succeed in climbing still higher, and *if these further losses are severe,* a good reason appears for the apparent redundancy of the early stages (i.e., of Figure 9.2): this redundancy is needed to create a reasonably large chance that at least one creature will succeed in the later stages. So we arrive at a picture not unlike that of John Bunyan's *Pilgrim's Progress.* Many set off in good heart along the road. Progressively they fall by the wayside, until at last only a few are left—at the end, in Bunyan's tale, indeed only one.

Let me make what I mean here more explicit by means of an example. In the midst of the crises we are now facing, we have begun to wonder how things may have fared with other creatures living on planets moving around other stars, and we have even begun to wonder about the possibility of communicating with such creatures.

Interstellar communication, as we may call it, raises many problems, some technical, some of general interest. Let it be said immediately that the only feasible mode of communication between creatures living on different planets moving around different stars would seem to be by radio. A vast array containing 900 individual radiotelescopes, each with a diameter of 100 meters, has been proposed. Such an array could actually be built now, and would be able to achieve interstellar communication. It has been named *Project Cyclops,* and is shown schematically in Figure 9.3.

The crucial question that we are led to face by the redundancy of the life forms of Figure 9.2 is this: Given a suitable planet, given the origin of life, given the emergence of intelligence to a level at least equal to our own, for how long on the average can we expect such an intelligence to persist? Even if intelligence arises on as many as a million planets in our galaxy, there will still be very few intelligent species around *at the present moment* unless high intelligence, once it arises, persists for more than 10,000 years, for the following reason. The age of our galaxy, the time span throughout which intelligence can emerge, is very long indeed—about 10,000 *million* years. Unless intelligence lasts once it arises, there will be very little overlap in time between its brief emergence on one planet and its emergence on another planet.

Suppose that our capacity to build an instrument of the technological quality of *Project Cyclops* might last for only 10,000 years. Is this an overly pessimistic assessment of the future of the human species? In view of the state of our present-day society, is it not rather an optimistic assessment? When one contemplates the huge human populations that have grown with startling suddenness during the last century or so, when one contemplates the excessive modern pressure on natural resources, it is hard to summon much confidence in a future extending more than a few decades. Devastating crises, one feels, must overtake the human species within a hundred years at most. We are living today, not on the brink of social disaster, as we often tend to think, but actually *within* the disaster itself. This is exactly what the news media report to us every day.

FIGURE 9.3
A schematic drawing of the massed radiotelescopes of *Project Cyclops.* (Courtesy of NASA.)

We have seen that the phenomenon of "intelligence" is an outcome of aggressiveness competition. Intelligence and aggressiveness are coupled together inevitably by the mechanisms of biological evolution. An intelligent animal anywhere in the galaxy must necessarily be an aggressive animal and must necessarily become faced at some stage by the same kind of social situation as that which now confronts the human species. Inevitably, then, "intelligence" contains within itself the seeds of its own destruction. Can any solution be found for this inherent difficulty?

In the next chapter we will consider this question, especially in terms of the conditions that would be needed here on Earth to enable our species to continue to maintain itself at a high technological level for more, or much more, than ten thousand years. A far smaller population would be needed, pressing only gently, if at all, on the resources of the Earth. It is hard to see our strident, competitive present-day society evolving smoothly in a more or less trouble-free way to the needed lower population level, or to see the persistently quarrelsome present-day human temperament changing voluntarily. To achieve such a change, psychologically as well as physically, an extensive selection of the human gene pool would probably be necessary. Some few individuals probably exist today with the necessary qualities, and it is from the progeny of these few individuals that the population of the future would have to come; the remainder of humanity, bearing the characteristics of our aggressive past, would have to become as extinct as the dinosaurs. In the violent future which lies ahead of us, these things may come to pass; yet I regard it as more probable that they will not. Inevitably, it seems the human species must then relapse back to its primitive condition. It seems that our moment of "intelligence," in a technological sense, will be exceedingly brief, that our ability to build and maintain *Project Cyclops* will not last for more than a century or two, perhaps for not more than the next few decades.

I see the uncertainties which now lie close for the human species as being an inevitable obstacle in the way of the emergence of any long-term intelligence. I see it as an obstacle every bit as formidable as the early physical problem of obtaining an appropriate planet moving around an appropriate star, and every bit as crucial as the origin of life itself. I suspect that many creatures may reach our present stage of development, but that only a few can go any further. Perhaps the chance of successfully surmounting the obstacle is as high as one in a hundred. Suppose each successful creature then has a life span of a hundred million years. On these reasonably favorable assumptions, the number of long-term intelligent species at present alive in our galaxy would not exceed about a hundred. It is among these fortunate ones that I would expect interstellar communication to be now taking place. The nearest of them would be unlikely to be less than 3,000 light years away from us.

It is to be observed that, for a species with a long-term future of a hundred million years ahead of it, a necessary interval of a few thousand years between the transmission of a message and the reception of a response to it would not seem a serious impediment. There would be ample time for many messages to be interchanged. For us, however, it is unlikely that much popular or political support will be forthcoming for *Project Cyclops* once it is understood that perhaps many centuries would be needed to obtain a positive result from it. Only if results could be promised in the short term would I expect such a project to receive public support. This I take to be clear evidence of the ephemeral nature of our modern society. We have no faith in tomorrow. A species with real confidence in its future would not hesitate to give expression to such a magnificent concept.

FURTHER READINGS

J. D. Bernal, *The Physical Basis of Life*. Routledge, 1951.

A. I. Oparin, *The Origin of Life on Earth*. Macmillan, 1938.

A. I. Oparin, *Genesis and Evolutionary Development of Life*. Academic Press, 1968.

10

EVERYMAN'S UNIVERSE

THOMAS ROBERT MALTHUS
(1766–1834)

(Photo from Historical
Pictures Service, Inc.)

10

EVERYMAN'S UNIVERSE

It would have seemed implausible only a few years ago to argue that the present-day state of society, and its future development, can be determined from energy considerations alone. Recent events, associated with what has become known as the "energy crisis," make this assertion seem less unlikely. In fact, an informed traveler from outer space could deduce the broad features of the present state of society, and its probable future, simply from the energy data given in Tables 10.1, 10.2, and 10.3.

Although energy plays a decisive role in our lives, there is no well-known unit of it. We buy gasoline and heating oil because of their energy content; yet quantities of gasoline and heating oil are specified by volume, not by energy. To make sure we have been supplied with the correct amount of energy, we then have to check that we have been given the right kind of stuff: gasoline, not water; heating oil, not cleaning fluid. We do this because the burners we use in practice are designed for fuels of a specific chemical nature.

TABLE 10.1
Amounts of energy per year from various sources

Source	Amount (in ergs)
Sunlight absorbed by Earth	3.4×10^{31}
Inanimate energy consumption by humans	about 2.5×10^{27}
Food eaten by humans	about 1.5×10^{26}
Muscular output of human species	about 3×10^{25}

TABLE 10.2
Energy available from various sources

Source	Amount (in ergs)
	Total
Deuterium in ocean waters (technology not yet available)	10^{38}
Uranium and thorium in continental granites (available with effort)[a]	10^{37}
Coal and Oil (available with modest effort)	10^{30}
	Per year (indefinitely)
Oceanic Tides (not feasible to obtain more than small fraction)	about 10^{31}
Wind (not feasible to obtain more than small fraction)	about 10^{29}
Hydroelectric (available with moderate effort)	2×10^{27}
Wood and plants (easily available)	10^{27}

[a] By "effort" here I mean cooperative effort among all humans, not one group competing economically with another. The present-day economic competition between nuclear power and power from coal and oil is a competition between a long-term energy solution and a short-term raid on irreplaceable resources. Such competition is highly deleterious to the survival of our species. A raid on the irreplaceable resources of coal and oil is justified only if it leads to access to long-term energy sources.

TABLE 10.3
Energy requirements for conducting various activities

Activity	Energy required (in ergs)
To extract uranium and thorium from low-grade sources	about 10^{35} (total)
To extract deuterium from ocean waters	about 10^{34} (total)
To extract energy of tides	too large to be practicable
To extract energy of winds	too large to be practicable
To cut coal and pump oil	small enough to be highly profitable
To set up hydroelectric installations	small enough to be profitable
To cut wood	small
To grow food	more than 1.5×10^{26} (per year)

If we replace heating oil by gasoline in the supply to the home boiler, there is an immediate disaster, and if we do just the opposite in the tank of our car, the motor refuses to run. Practical considerations of this sort prevent us from buying fuels by their direct energy content, with the consequence that we have only a rather hazy concept of what "energy" really is.

The situation is made worse by a confusion between power and energy. Electric devices are given power ratings, usually in watts or kilowatts (1 kilowatt = 1000 watts). The electrical energy consumed by such devices is calculated from the equation

$$\text{Electrical energy consumed} = (\text{power rating}) \times (\text{time used}).$$

A 100-watt device used for 100 hours consumes the same electrical energy as a 1-kilowatt device used for 10 hours. The bill from the electric company will be the same in the two cases, so once again the quantity we buy from the electric company is energy, not power. A 1 kilowatt device used for 1 second consumes a quantity of energy known as the *kilowatt-second.* This quantity can be taken as an energy unit, although in physics it is more usual to use a much smaller unit known as the *erg.* The erg is related to the kilowatt-second by the equation

$$1 \text{ kilowatt second} = 10^{10} \text{ ergs.}$$

The erg has been used as the unit in Tables 10.1, 10.2, and 10.3.

Let us suppose our traveler from outer space knows the size and mass of the Earth, the distance of the Earth from the Sun, and some other related astronomical and geophysical details. Observing the annual inanimate energy consumption of the human species, the traveler would be able, by the following reasoning, to infer the approximate level of our technology. First, from the fact that the inanimate energy consumption exceeds by an order of magnitude the energy content of the food eaten by humans, it would be clear that some form of industrial technology had been achieved. To achieve industrial technology, a brain of adequate capacity is needed, requiring the human to be not a small creature, like a mouse, but large enough to support a brain of the appropriate capacity; then, knowing the strength of terrestrial gravity, our traveler would be able to make a tolerable guess as to our individual size and weight. This would define the muscular output of a single individual. If the traveler knew the total muscular output of the whole species, the approximate number of humans could be deduced (actually about 4 billion). Finally, dividing the total inanimate energy consumption by the number of humans, the traveler would arrive at a figure for the average energy consumption per head. This figure would decide the level of our technology, since technological sophistication is closely correlated with per capita energy consumption. Human society today has a per capita energy consumption about ten times greater than the per capita food consumption. It is this factor of ten which creates the difference between modern industrial civilization and the civilizations of Greece and Rome—in the latter, inanimate energy was about the same as food energy. And, in an opposite direction, a civilization with a thousandfold excess of inanimate energy over food energy would inevitably be far more technologically sophisticated than we are today.

Actually, our traveler would also be likely to make some quite serious mistakes. Given only energy values for the whole human species, the traveler might naturally assume that the per capita consumption of inanimate energy is about the same everywhere over the whole Earth. The disparity between developed countries and the more populous underdeveloped countries would probably seem too unlikely a situation to be considered seriously. An average inanimate energy consumption equivalent to the burning of only about two tons of coal per person per year (less than the average monthly consumption in the U.S.) would suggest erroneous conclusions. From considering heating problems alone, the traveler would infer either that few humans live outside the tropics or else that we are furry animals. The traveler would argue that, to maintain an industrial society on a per capita annual consumption as small as two tons of coal, we must be extremely careful not to permit waste. There could be comparatively little traveling from place to place. Each person would need to live close by his work. The existence of energy-wasting automobiles or jet planes would not be contemplated. Perhaps most erroneously of all, the traveler would think that our society must be carefully balanced, conservationist minded, rather rigid in its social structure, and consequently much more stable than it actually is. Technological progress would in such circumstances be slow and rather ponderous, not at all like the potentially explosive technological capability that we actually possess. Nor would our traveler appreciate how close we are to the collapse of our present civilization.

A good deal of doomsday talk is abroad nowadays, concerning the problems of a continuing supply of energy and of reserves of metallic ores. Although this talk contains a very serious point, most of it is incorrect. There is no intrinsic shortage facing the human species, either of energy, of food, or of metallic ores. Sufficient energy from wood, plants, and hydroelectric sources is available to maintain us at our present level of technological sophistication for an indefinitely long period. Nuclear energy is potentially available on an enormous scale. The deuterium* in the ocean waters could yield sufficient energy to maintain a technological level far above our present level for millions of years, if we could figure out how to fuse hydrogen into helium.

---

*The nucleus of the deuterium atom is the deuteron (mentioned on page 70), composed of a neutron and a proton.

As we saw in Chapter 4, the sun is a nuclear furnace, producing its energy by converting atoms of hydrogen into atoms of helium, four atoms of hydrogen being fused together to make one atom of helium. The processes whereby this occurs have been studied by physicists. They can be reproduced on the Earth (in the hydrogen bomb), but not so far by any process that would be useful on a commercial scale. In order to be able to produce energy in useful amounts in this way, we would need to raise hydrogen gas to a very high temperature and *to maintain it there.* The trouble on the Earth is that, whenever we make a gas very hot indeed, it simply blows apart in an explosion. The hot gases inside the Sun do not blow apart because they are held together by gravity. The Sun contains so much material that gravity within it is extremely strong. On the Earth it is not possible to reproduce this condition, because the gravity created by even the mass of the whole Earth is not strong enough to restrain gases that are as hot as they are inside the Sun. Instead of gravity, scientists have tried to use magnetic forces (and deuterium rather than hydrogen) to solve this problem, but this has turned out to be difficult. (Governments throughout the world sponsor efforts along these lines, for the obvious reason that the energy yield from the conversion of hydrogen to helium would be enormous, if only a way could be found to do it successfully. The hydrogen in a few thousand tons of ordinary water would, by being converted to helium, provide enough energy to meet the needs of the whole of the world's industry for a year.)

So what the Sun does very easily, man still finds very difficult. A much simpler way of obtaining energy is to use natural sunlight as it falls on Earth. The sunlight falling on the roof of a house gives sufficient energy for the needs of the household, but there is the difficulty that the heat of summer must be stored for use in winter. Modern science offers nothing better than storage in expensive and cumbersome electric batteries, which are much too inefficient to be considered a satisfactory solution. The best way to make use of summer heat in winter is still the age-old method of burning in winter the wood that was grown in the summer.

A less direct way of using sunlight is by means of the water that evaporates from the oceans. The water vapor rises from the ocean surface to considerable heights in the atmosphere. Some of it falls as rain and snow on high ground, where it can be trapped in lakes. The water from such high lakes can then be run downhill and made to drive electrical machinery. This hydroelectric power would be adequate to run the world's industry if the human population were, say, about a half of its present size.

In the 1950's there was a euphoria among scientists and politicians alike concerning fusion as a source of energy, which was represented as being "just around the corner." The euphoria was at that time misplaced, but the situation has changed markedly in recent years. Ideas are being considered today which have a sensible look about them. It begins to look as though the quite fantastic energy potential of deuterium may one day be realized. But not within only a few years. We all need to realize that obtaining energy from deuterium is a project for the future, not for our own day and age. The position today is somewhat similar to that when Michael Faraday invented the electric motor. Faraday showed that ingenious devices were physically possible, but decades would pass before they became important practical developments.

And even if obtaining energy from deuterium does not eventually become a feasible project, we can certainly get energy from uranium and thorium. Table 10.2 shows that the energy content of the uranium and thorium in the crystal rocks of the Earth is only about one order of magnitude smaller than that of the deuterium in the ocean waters, equally enormous when judged by present-day standards.

Then why is so much doubt expressed nowadays about energy availability? Because the underlying issue is whether energy can be made available fast enough and easily enough to avoid significant perturbations in the life style of the middle-class citizens of the U.S.A. and of Europe. If such perturbations be interpreted as a doomsday situation, then indeed the present state of affairs is unsatisfactory. Even the widespread use of hydroelectric energy would be less convenient* than our present use of coal and oil, and the widespread use of nuclear energy would raise social problems that are new, strange, and, to many people, uncomfortable. What the doomsday prophets are actually concerned about is the possible change in our life style, not the physical exhaustion of the energy supply. Energy *cannot* be exhausted. A particular mode of access to energy can become exhausted, however. For example, Table 10.2 shows that coal and oil will last for only about 400 years.** Even the most rigorous attention to conservation could only extend the 400 years to 500 years, or 600 years, or 1,000 years. The ultimate outcome is inevitable. Coal and oil cannot be used in economically significant quantities without becoming exhausted in a time-scale appreciably less than the span of written history. This circumstance is so self-evident that I sometimes wonder just what case the conservationists think they are arguing. There is no special merit in arranging a situation in which coal and oil last for 800 years rather than 400 years—the difference is wholly insignificant compared with the biological evolutionary scale. The really relevant issue is whether coal and oil will last long enough to enable us to achieve access to an alternative energy reservoir, one that would be, in effect, inexhaustible.

The concept that metallic ores can be exhausted is also erroneous. A given ore deposit may become exhausted, and all deposits having more than a certain specified grade may become exhausted. But there are always new deposits of lower metal content available.

*At least half the world resource of hydroelectric energy is in South America and Africa, far removed from present centers of technology and industrial production.

**If *all* the coal that exists could be recovered at zero energy cost, this estimate would be increased to more than 1,000 years. But the recovery of much of this coal would require expending as much energy as the coal would deliver (using foreseeable mining techniques).

The lower the ore grade, the larger the quantity of it that can be found—if x per cent is the metal content, the amount of ore which can be found seems to vary about as $1/x^3$. It is true that more energy, and a more refined technology, is needed to extract pure metal from a low-grade ore than from a high-grade ore, but these are practical issues, not questions of principle.

Having thus disposed of the less serious threat to our society, let us begin to consider the more serious issues. A primitive society of individual farmers, or of small groups of hunters, is not exposed to the inherent instability which afflicts large interdependent groups of people. In ancient times bad weather could indeed lead to difficulty, perhaps to death by starvation for particular individuals, but such vicissitudes did not occur simultaneously everywhere—bad weather in one place was usually offset by better weather somewhere else. But once the first river-valley civilizations were established in Sumeria some 6,000 years ago, a tight connection became established between the organization of society and its productivity. Productivity, initially of food, was lifted significantly by the organizational structure itself. With the increase of food supplies, the population then increased beyond the size possible for a society of more or less disconnected individuals. The human organization, relating the work of each person to the work of others, became necessary simply to maintain the *status quo*. The instability which afflicts all civilizations was thereby established.

Perturbation of the social organization would change productivity in a way that reinforced the effect of the initial perturbation. In short, a positive-feedback situation became established, with organization and productivity forming a looped relationship,

$$\text{Social organization} \rightleftarrows \text{Productivity.}$$

The positive feedback here can work either to amplify the relationship or to diminish it, depending on the nature of the initial perturbation. Perturbations in the early civilizations were of a rather simple kind: war between rival cities, rivalries between important families.

Throughout written history there has generally been an increase in the complexity of the organization of society, which might be taken to show that the looped relationship almost always works in an amplifying sense. However, I believe this comfortable conclusion to be in error. In a world of many communities, those which happen to undergo amplification are able to overwhelm and swallow up those which experience attenuation. Indeed, if only one community among many experiences an amplifying loop, it will be this one particular community that dominates the writings of historians. So we shall be misled into imagining that amplification occurs inevitably all the time. So far from this being true for any one particular community, I suspect that a particular community, if unaffected from without, is just as likely to experience attenuation as amplification.

The increase of productivity of a single community can sometimes get ahead of the rise of its population, because major advances in productive techniques tend to go in steps, as in Figure 10.1, whereas the rise of population goes in a smooth curve, as in Figure 10.2. The reason for the difference between the curves is that major advances in techniques often come from ideas which happen suddenly (or a great new tract of land may suddenly become available for production, as happened with the discovery of the Americas), whereas the rise of population takes time—time is needed for children to be born and for them to grow to maturity. Combining Figures 10.1 and 10.2, we have the situation of Figure 10.3, in which the productivity curve sometimes rises above the population curve. It is in these steplike rises that amplification cycles occur. But given time, the population curve always tends to rise above the productivity curve. Only then does the population

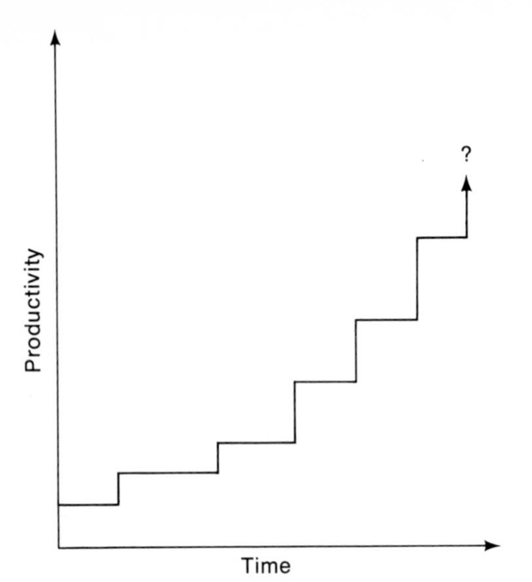

FIGURE 10.1
Major advances of productive techniques tend to occur in steps.

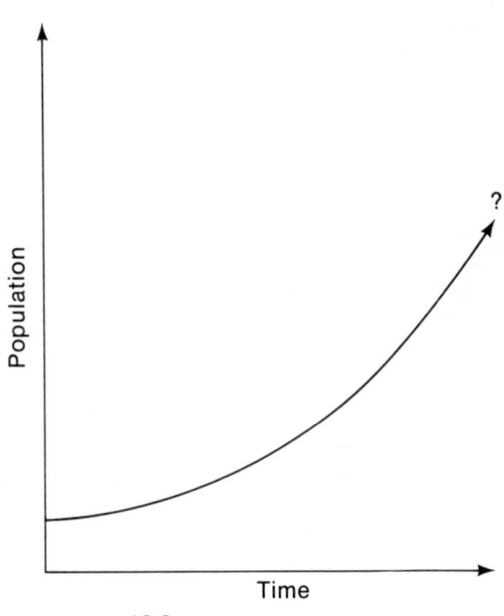

FIGURE 10.2
A rising population follows a smooth curve, because time is needed for children to grow to maturity.

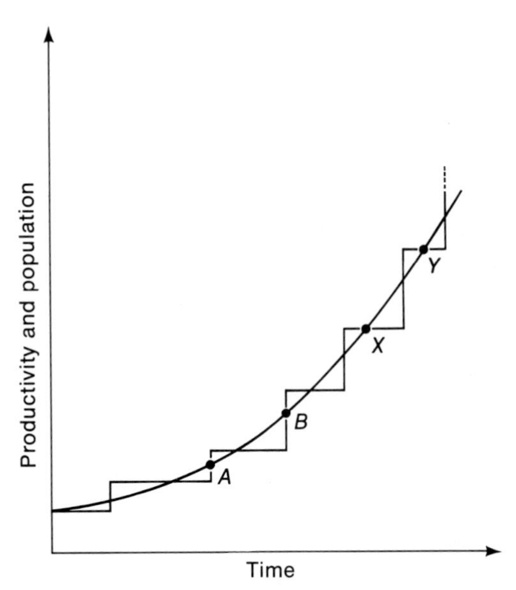

FIGURE 10.3
In the interrelation of productivity and population, points *A* and *B* favor expansion, whereas *X* and *Y* are likely contraction points.

curve become moderated by shortages. Unless a further jump of productivity occurs, the community in question is likely in such a situation to enter an attenuation cycle. The points *A* and *B* of Figure 10.3 are likely amplification stages, and points *X* and *Y* are likely attenuation stages.

It seems probable that all human beings will soon be living in what is effectively a single worldwide economic community. Applying the preceding ideas to such a community, can we expect jumps of productivity always to keep ahead of the rising population curve? To answer this obviously critical question, we note that productivity depends on energy availability. Although the energy available to the human species may eventually rise appreciably above the present rate of about 2.5×10^{27} ergs per year, there is a limit, probably in the region of 10^{29} to 10^{30} ergs per year, which must eventually be reached. There is no mathematical limit, on the other hand, to the potential rise of population. If the world population were to continue increasing at its present rate of about 2 per cent a year, the number of people alive in 1,500 years would be so large that the whole of the Earth's atmosphere would only suffice for a single breath of air for each person. And in less than 5,000 years the mass of the human species would exceed the combined mass of all the galaxies visible in the largest telescopes. Rising population can always exceed any specified level of productivity, so that points like *X* and *Y* of Figure 10.3 always exist, regardless of our technical ability and productivity.

A typical population-development curve for an isolated community might look like Figure 10.4. The over-all development is not unlike the lift of a wave, which, after rearing up, breaks and turns back to a lower level. The eventual collapse of the social organization causes productivity to be cut back, with the population then falling to a level where it can be maintained by a weaker technology.

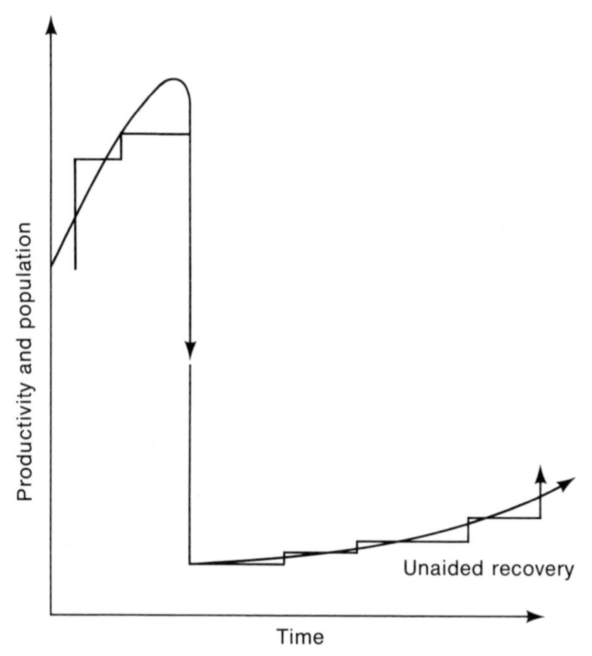

FIGURE 10.4
The collapse of an isolated community, followed by a slow unaided recovery. The more advanced the technology of a society, the greater the collapse is likely to be.

Although no community has yet existed in isolation from all other communities, a rough approximation to the evolution of a self-determining community is to be found in the rise and fall of the Roman Empire. The early rise of Rome was associated with technological advances in agriculture and road building, with the use of aqueducts to obtain good water from the Alban hills, and with the disciplined organization of the Roman army. The later rise was caused mainly by territorial expansion. After the breaking of the Roman wave, there could be no quick recovery by interaction with other communities, since there were no other suitable communities in adequate contact with the Roman lands. Rome had spread so widely, was so much on its own, that recovery, when at last it came, had to be spontaneous. The long period of recovery is often described by historians as a "Dark Age." Such an age is not as unpleasant to live in as the description "dark" might suggest. Historians call the period which followed the collapse of the Roman Empire "dark" merely because we have so few surviving documents from the period that historians lose sight of the details of what went on then (from about A.D. 450 to 900). That is, the period is "dark" to us, not to those who lived in it. Actually, it was a period of extremely fruitful invention, in which Roman technology was not only recovered or maintained, but even much improved upon. Many new discoveries were made during this extended recovery phase, discoveries which then served as a springboard for the next rising wave of the world population, the wave which is our own. On the whole, such recovery periods are good times to live in, better than the rigid, fossilized state of affairs which is likely to exist just before the break of the wave.

In its broad features, the rise and fall of Rome is a blueprint for what is likely to happen to our present-day society. A reasonable estimate can even be given for the time at which our particular wave will break. Within an error of perhaps a decade or two, the year will be 2025.* It is of interest to trace the developments that seem probable for the next fifty years, from the present to the break of the wave.

A rapid evolution toward a single worldwide community is now operative. There are three broad groupings of humanity: the western set of developed nations (but including Japan, Australia, and New

*The rate at which the world population is *now* increasing is about 2 per cent per year. *But the annual rate of increase is itself increasing.* If the rate of increase were to behave in the future as it has behaved during the past three or four decades, the world population would tend to infinity in about the year 2025. Of course, the number of people cannot actually become infinite; a crisis must intervene, in the form of a collapse of the world culture that is now producing this grotesque situation.

Zealand); the communist bloc of China, Russia, and their dependents; and, third, the very heavily populated underdeveloped nations. No stable worldwide community seems possible as long as the present large disparity between living conditions in these three groups continues to exist. The western group is under more pressure to level up with the underdeveloped group than the communist bloc is. The concepts of democracy, individual freedom, human rights, all expose the open societies of the west far more intensely to pressure from demands that they share their resources with the underdeveloped group than does the philosophy of the communist bloc, in which countries the voices of individuals are fiercely muted and the state is all powerful.

The western nations are peculiarly defenseless against this pressure, because it is generated internally, a circumstance which arises from the nature of western democracy, specifically, from our method of democratic elections, which causes political leaders to attach overriding importance to the winning of elections.* The western leaders are therefore concerned with gaining the approval of a majority of the community, and if the majority happens to be lazy rather than industrious (perhaps because of indulgence in childhood), the leaders inevitably seek to appeal to the lazy rather than to the industrious. That is, they legislate for a leveling-up process, by taxation or by "nationalizing" capital.

This trend does not stop at the geographical boundary of a nation. It is carried over into the foreign policy of a nation, and if not just one nation but many are so affected, the concept of a worldwide leveling-up process gains momentum. This is precisely what has happened during the past quarter-century. The concept of

*Those who do not rarely win elections. Nevertheless, it is from the few unusual cases, where a person sets other considerations ahead of winning and yet happens to win, that outstanding statesmen emerge.

leveling up has come from within the western group of nations itself, accompanied by the implausible suggestion that leveling up will itself check the rise of world populations, a suggestion which seems absurd in the light of the facts expressed in Figure 10.5. It is just the leveling up that has taken place already which has permitted so many of these curves to rise so high.

At first, the worldwide application of the leveling-up idea gained influence for the western nations, particularly for the United States. Twenty years ago the U.S. could win any vote it chose to win at the United Nations. Today the U.S. can scarcely win any vote at all, so rapidly has the momentum developed. The democratic western system of one-man-one-vote is now being translated into an international one-member-one-vote at the United Nations. There is, of course, no need for the U.S. and the other western nations to accept such a way of deciding world policy, but because of our own concepts and ideals, it seems likely that we shall, and even more so in the future.

The effect of a leveling up between the west and the underdeveloped nations will increase about fourfold the population load on western productivity, with the essentially certain result that the rising wave of western society will break, as in Figure 10.4. If for the moment we imagine the communist bloc not to exist, the population level would fall to a lower level, where it could be supported by a less sophisticated technology. The rise and fall of the Roman Empire would have been repeated. There would then be a new Dark Age, out of which a new wave would quite likely arise.

The actual presence of the communist bloc creates a complication. Two variations have to be considered. The collapse of the western wave may carry the communist bloc down too, in which case the pattern would be fundamentally the same. In the second, more interesting variation, the communist bloc may turn out to be strong enough to produce an early rise once the western wave has

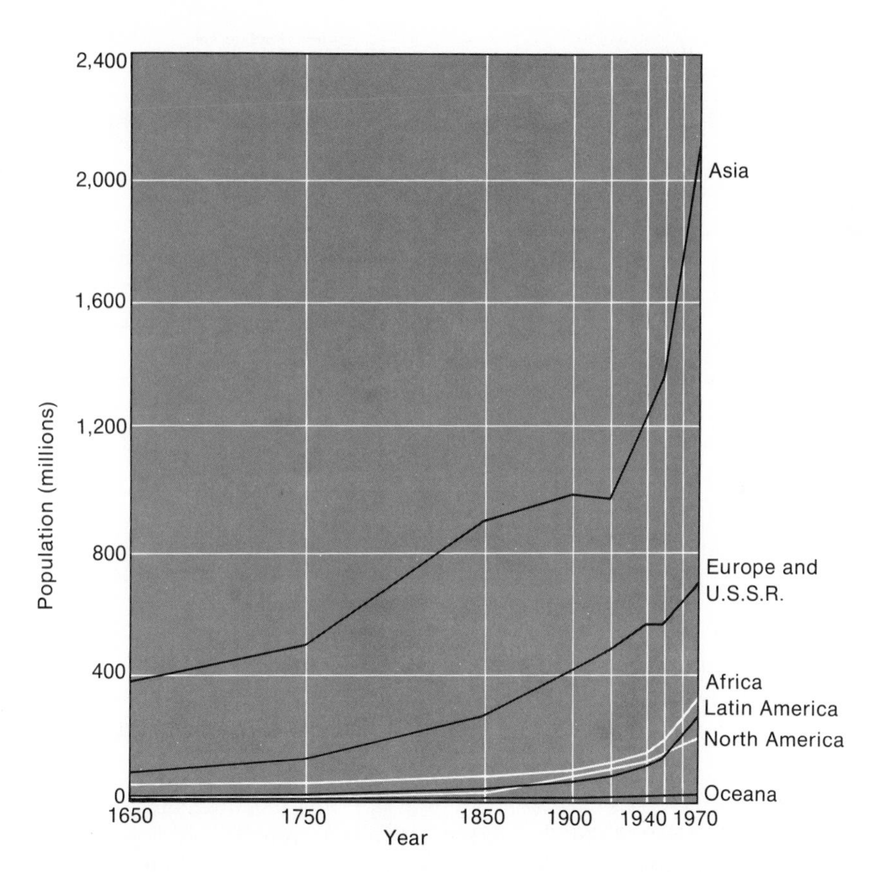

FIGURE 10.5
Regional growth of the world's population, from 1650 to 1970. In 1650 total world population was about 553 million. By 1850 it was 1.3 billion, and in 1970 was 3.6 billion. (From T. Frejka, "The Prospects for a Stationary World Population." Copyright © 1973 by Scientific American, Inc. All rights reserved.)

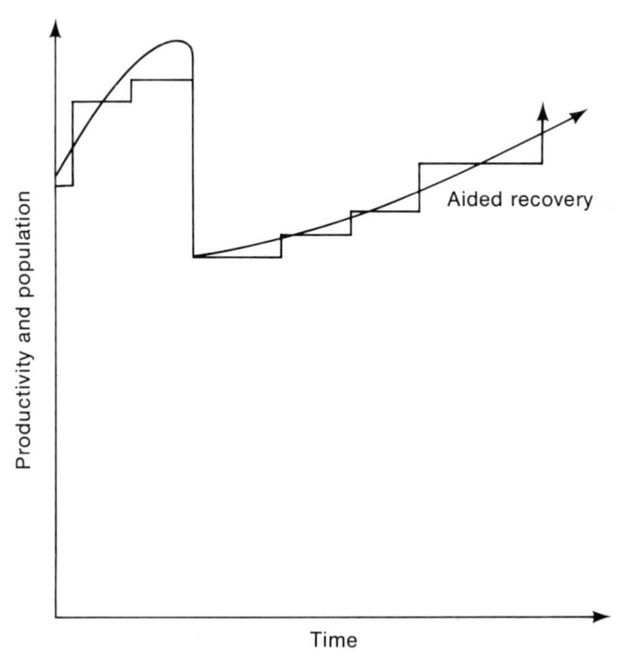

FIGURE 10.6
Recovery of the west, promoted by help from the communist bloc.

fallen, as in Figure 10.6. In this second case we have but a temporary dip following the western collapse, simply a hiatus in the rise of a world society, one that now would inevitably be dominated by the philosophy and ideas of the communists. These are ideas which deny the concept of individual freedom, of one-man-one-vote (except in the peculiar one-party elections practised by the Russians), and which would consequently deny the worldwide operation of the concept of one-nation-one-vote. Such a society might be better able to prevent the eventual break of the wave of the worldwide community, particularly because the denial of individual freedom might permit the denial of the freedom which has hitherto caused the eventual collapse of every society: of the early civilizations of the Near East, of Greece and Rome, and now, indirectly, through the underdeveloped nations, of the whole of the west.

The freedom in question is, of course, the freedom to procreate. All manner of other freedoms have been given up from time to time—freedom of speech, freedom from arbitrarily heavy taxation—but never the freedom to procreate. Yet, ironically, this freedom is not as critical to the individual as are many of the freedoms which are taken away from us by the unchecked rise of population. Today, in Britain, the birth rate has fallen to replacement rate, and this has been done voluntarily. The same situation is close to hand in much of Europe, and is even within sighting distance in the United States. However, such voluntary limitation would not work for the whole world, since it would leave the way open for the operation of a powerful biological principle.

If among several varieties of a species there is just one that rapidly increases in number, that increasing variety will eventually overwhelm all the others, and eventually all surviving members of the species will belong to the expanding variety. In short, the expanding variety becomes the cuckoo-in-the-nest of the whole species. Applying this principle to our own species, if any one group of humanity refused to honor a voluntary agreement on family limitation, all other groups would ultimately be eliminated. Obviously, voluntary agreements would therefore be open to general suspicion, and so would be neither acceptable nor workable. Indeed, Figure 10.5 shows that certain rapidly increasing populations are already well on the way toward biological elimination of all other human groups.

It is customary to regard these rapidly expanding groups with sympathy, since the people involved are for the most part very poor. Their poverty is in large measure a consequence of the population expansion, however, and in fact it is physically possible for the catastrophic rise of the curves of Figure 10.5 *to be checked and brought under control within not much more than six months.** The

*By adjusting the birth rate to equal the death rate. This can lead to maladjustment of the different age groups, but in a clearly desperate situation this self-caused difficulty must be faced.

fact that the population explosion continues unabated from decade to decade shows that the groups in question have no collective wish to moderate their behavior. Because of the long term durability of its effects, unrestricted expansion by a particular group is really a profound form of aggression, more drastic than the aggression of an invading army. For a slowly expanding group to treat this biological aggression with sympathy is to make its own extinction all the more certain.

We are now in a position to return to the problem reached at the end of the preceding chapter, the problem of the long-term survival of the human species. Since at this moment we do not know whether the freedom of the individual to procreate will or will not be curtailed, we must discuss the problem in terms of both these possibilities.

First, suppose that no restriction is applied to the freedom to procreate. Then nothing will prevent our current wave of productivity and population from eventually breaking in the manner of Figure 10.4, probably around the year 2025, as I mentioned earlier. A succeeding Dark Age will follow, possibly lasting for several centuries. Like the Dark Age that followed the collapse of the Roman Empire, this new Dark Age will probably be an age of invention, an age with more individual freedom than we experience today, an age that is rather quiet and pleasant to live in. Relatively easily accessible supplies of oil and coal and of mineral deposits will still be available beyond the year 2025. Energy from wood and hydroelectric sources will still be available. Books and other documentary information from our present civilization will no doubt survive; so technological processes will not have to be invented *de novo*. There is accordingly no reason why, after a few centuries, a further wave of productivity and of population should not arise. Yet nothing will be different for this further wave. It too must eventually collapse, for just the reasons we have already discussed.

In this way, assuming no restrictions on population growth are applied, we arrive at the concept of Figure 10.7. Like ocean waves rolling into a beach, wave after wave of human society will rise and break, perhaps with a cycle time of about 500 years. Yet the form of Figure 10.7 cannot continue indefinitely into the future. Just as the heavy surf following a storm eventually subsides, so the waves of Figure 10.7 will subside, as in Figure 10.8. When this stage is reached the "experiment" of intelligent life on the Earth can be considered to be at an end. Mankind would still continue to exist, but for all future time at a primitive level. No machines would be available for transportation, for the growing of food, or for producing manufactures. The situation would be more primitive than any state of society experienced within the last 6,000 years. The reason for this final decay of civilized society lies in the exhaustion of readily accessible supplies of raw materials. Oil and coal will go after but a few cycles of Figure 10.7. Although simple energy sources like plants and wood, and perhaps hydroelectric energy, will remain available, metal technology must collapse eventually. Continued access to metals requires ores of lower and lower grades to be worked. Although this requirement would not present a serious problem to a society with steadily improving technology, it must eventually present an insoluble problem to the Dark Ages which follow the collapse of the successive waves of Figure 10.7. The waves will then die away in the manner of Figure 10.8.

It is equally straightforward to foresee the situation that would follow the limitation of the population early enough during one of the uptrends of Figure 10.7. With the productivity curve then rising above the population curve, in the manner of Figure 10.9, ample effort for further advances would be available. As the productivity curve rose, giving more and more investment possibilities for the development of new ideas, the way would be directly open to the enormous energy resources given at the head of Table 10.2. With

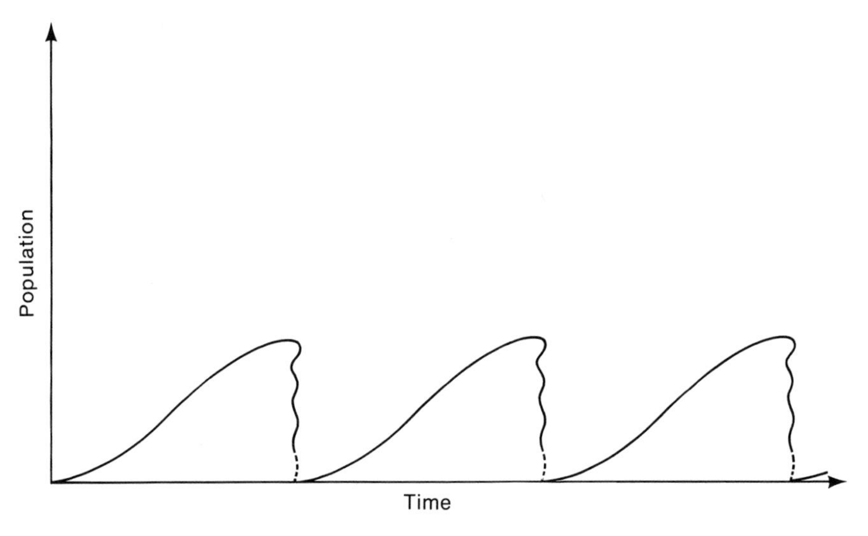

FIGURE 10.7
A schematic representation, assuming no compulsory population limitation, of the breaking of successive waves of human society.

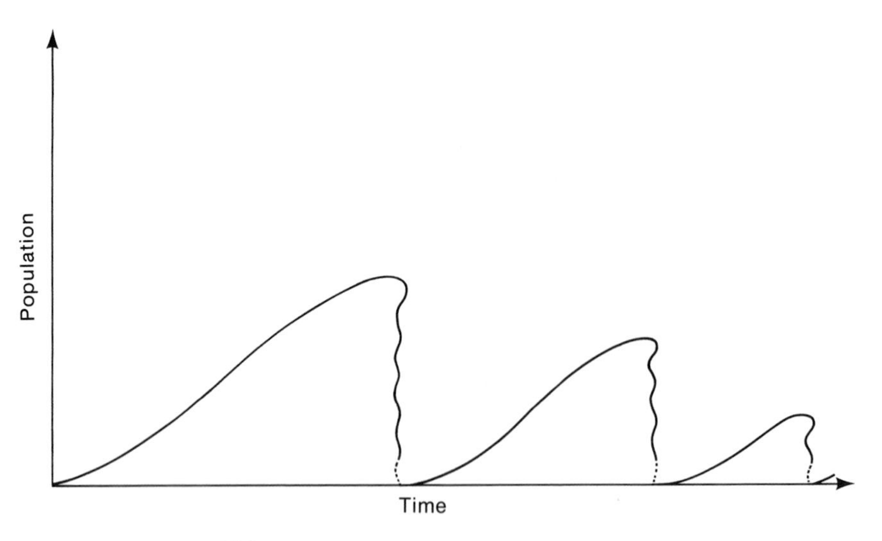

FIGURE 10.8
Assuming no compulsory population limitation, the successive waves of Figure 10.7 die gradually away, as easily accessible supplies of energy and of minerals become exhausted.

energy availability climbing far beyond the present-day level of Table 10.1, the way would be clear for the human species to emerge as one of the creatures who "make it" in the sense of the preceding chapter. Our species might then look forward to a future measured not in centuries, but in millions of years, a future in which our descendants would look back on us in much the way that we look back on our remote ancestors of a hundred thousand years ago. Our descendants would think of us as unfortunate creatures, ill-equipped emotionally and psychologically to overcome the social problems of our time, just as we regard our ancestors as being ill-equipped technically—ill-equipped to deal with the rigors of an ice age and seemingly ill-equipped to deal with formidably armoured animals, the bear, and the great cats. I would expect the intellectual horizons of our descendants to become altogether wider than anything of which we can conceive today, just as our present-day horizons lie beyond the range of stoneage man.

An attempt, on the other hand, to limit the population level near the crest of one of the waves of Figure 10.7 would, I believe, be fraught with great danger, because such an attempt would be half-hearted. It would run counter to the emotional structure of society, and would therefore tend to be abandoned should a further rise of productivity occur. The stable situation of Figure 10.9 would be unlikely to arise. Instead there would be the stop-go situation of Figure 10.10, with temporary population limitations occurring from time to time, but with the population always overtaking the productivity curve. The danger in such a partial restraint lies in the wave being driven to very great height before at last it breaks. The collapse would then be catastrophic, with the ensuing Dark Age so truly dark that nothing at all could emerge from it.

The danger here is well-illustrated by our present-day situation. The steps that would be required to give an effective and permanent limitation to population growth are deeply inimical to our

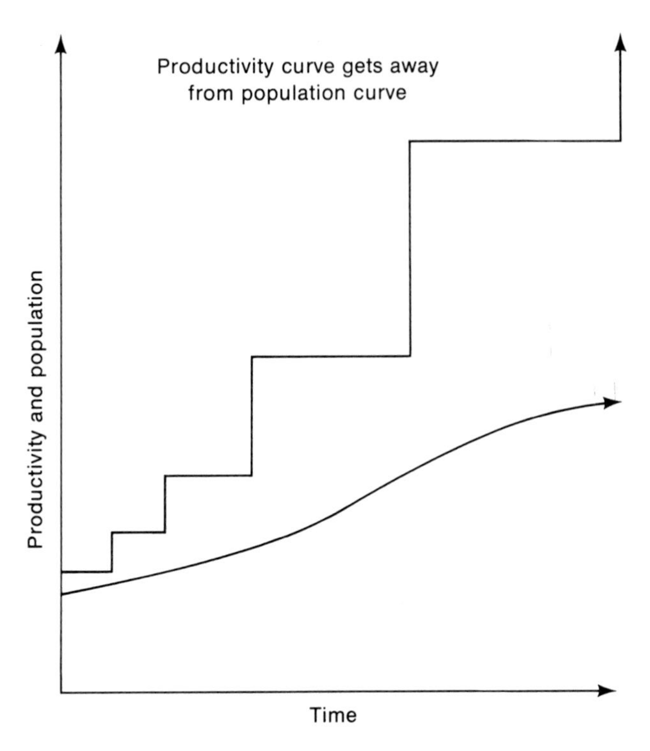

FIGURE 10.9
Assuming compulsory population limitation, productivity can advance far beyond the material demand of society, permitting enormous investment in new techniques and ideas. In this situation, humanity becomes one of the creatures in the Galaxy that "make it" through to a higher level of the hierarchy of life.

present-day global culture. It is difficult to imagine an effective rule that would be milder than one in which the parent of a third child becomes subject to taxes that would be crushing enough to provide a serious restraint. Nothing weaker than this would have the remotest chance of proving effective. It is true that temporarily adverse circumstances can lead to voluntary population restraint, but rises of productivity will then always promote a further population expansion. This is exactly what must not happen if the situation of Figure 10.9 is to be achieved. Clearly a society with different cultural values than our own would be better able to

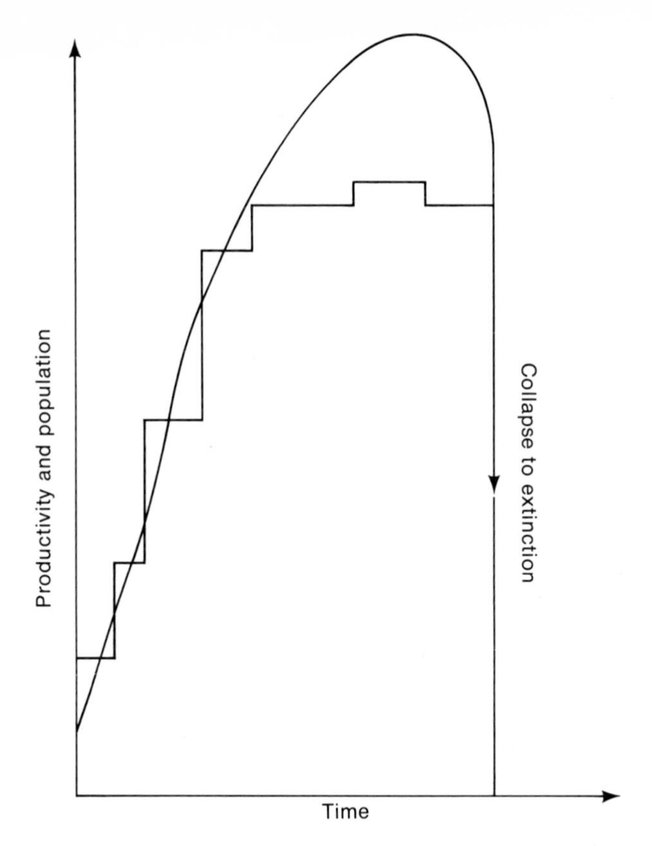

FIGURE 10.10
A continuing stop-go relationship between population and productivity can lead to a collapse of so great a magnitude that no subsequent recovery occurs. This is the road to extinction for the human species.

accept the kind of rule which is needed if the fate of our species illustrated by Figure 10.8 or by Figure 10.10 is to be avoided.

I come finally to the most ominous circumstance of all. The estimated date of 2025 for the collapse of our society is close to the time at which ample energy from the fusion of deuterium may become technically feasible. Should very large energy supplies become available *before* the present wave breaks, the moment of the break might well be long postponed. Both productivity and population would then rear up higher and higher in a wave of great height, as in Figure 10.10; the final crash might then lead to the extinction of our species—in the total breakup of a society given to handling large quantities of radioactive wastes, extinction would not be unlikely.*

Nuclear energy, although essential to a major advance in the manner of Figure 10.9, therefore also carries with it the danger of ultimate disaster. The issue turns on whether population is limited before the technical problems have been solved. Of all the issues that have ever faced our species, this appears to me the most critical. No other technological step has ever at any time carried with it these opposing possibilities, of an essentially unlimited advance or of total extinction.

Turn back to Figure 10.7. Since the highest points of the successive waves are the epochs of highest technological competence, it is just at one or another of these wave crests that the problems of nuclear energy are most likely to be solved. Yet it is just at these places where the disastrous situation of Figure 10.10 is most likely to develop. One must therefore conclude that the chance that the human species will survive is rather small, even in the attenuated and unsatisfactory form of Figure 10.8. Only if a cultural pattern leading to rigorous population control occurs *early* in the rise of a wave, and only if nuclear energy becomes readily accessible later in that particular wave, can our species expect to emerge into the favorable circumstances of Figure 10.9.

In conclusion, then: the danger facing our species, and arising from the rapid increase in world population, will reach a crisis point around the year 2025. If an effective nuclear technology is set up *before* collapse occurs, then extinction is likely. If collapse occurs

*The same argument would also apply if energy becomes easily available from uranium and thorium. The real danger of nuclear energy lies in permitting a further larger increase of the world population, not in the arguments currently being waged by environmentalists.

first, on the other hand, then a favorable opportunity will occur a few centuries from now, at the beginning of emergence from the ensuing Dark Age. Should population limitation then occur and be followed by the discovery of deuterium fusion or by the provision of essentially unlimited energy from uranium and thorium, the next wave of human effort could continue to rise, without break, into a new civilization.

We humans carry with us the heritage of a long past, extending backward in time for hundreds of millions of years. Our society is built not on the joy and happiness of the past, but on the agonies experienced by the long line of our predecessors. Whether or not all the agonies and struggles of the past will emerge into a great future, or will vanish into nothing at all, is likely to be decided in the next few tens of human generations.

FURTHER READINGS

H. A. Bethe, "The Necessity of Fission Power," *Scientific American*, **234** (January 1976), 21.

S. E. Hunt, *Fission, Fusion, and the Energy Crisis*. Pergamon Press, 1974.

INDEX